NATURAL DISASTERS: HURRICANES

A Reference Handbook

Other Titles in ABC-CLIO's
**Contemporary
World Issues**
Series

Books in the Contemporary World Issues series address vital issues in today's society such as terrorism, sexual harassment, homelessness, AIDS, gambling, animal rights, and air pollution. Written by professional writers, scholars, and nonacademic experts, these books are authoritative, clearly written, up-to-date, and objective. They provide a good starting point for research by high school and college students, scholars, and general readers as well as by legislators, businesspeople, activists, and others.

Each book, carefully organized and easy to use, contains an overview of the subject, a detailed chronology, biographical sketches, facts and data and/or documents and other primary-source material, a directory of organizations and agencies, annotated lists of print and nonprint resources, a glossary, and an index.

Readers of books in the Contemporary World Issues series will find the information they need in order to better understand the social, political, environmental, and economic issues facing the world today.

NATURAL DISASTERS: HURRICANES

A Reference Handbook

Patrick J. Fitzpatrick

CONTEMPORARY WORLD ISSUES

ABC-CLIO

Santa Barbara, California
Denver, Colorado
Oxford, England

Library of Congress Cataloging-in-Publication Data
Fitzpatrick, Patrick J.
   Natural disasters : a reference handbook / Patrick J. Fitzpatrick.
      p.   cm.—(Contemporary world issues)
   Includes bibliographical references and index.
   ISBN 1-57607-071-9 (alk. paper)
   1. Hurricanes.   I. Title.   II. Series.
QC944.F58   1999
551.55'2—dc21   99-40280
                                                                      CIP

04   03   02   01   00   99   10   9   8   7   6   5   4   3   2   1

ABC-CLIO, Inc.
130 Cremona Drive, P.O. Box 1911
Santa Barbara, California 93116-1911

This book is printed on acid-free paper ∞.

Manufactured in the United States of America

Dedicated to the victims
of Hurricane Georges, Hurricane Mitch,
and Hurricane Floyd

Contents

Preface

The year this book is being written, 1998, marked both an exciting and cataclysmic year involving hurricanes. This year ushered in a new era of unprecedented hurricane observations that will improve our understanding of these storms. In addition, new technologies are emerging that will provide even more detail about hurricanes in the future. Ultimately, these advances will improve hurricane forecasts. In 1998 we witnessed:

- The most comprehensive hurricane measurements to date as a result of collaborations between NASA, the National Oceanic and Atmospheric Administration (NOAA), and the U.S. Air Force Reserve. These missions, which included up to six planes at once in a hurricane, measured temperature, wind, and moisture from the earth's surface to 40,000 feet. Remote sensing technology on the aircraft and on satellites in space provided information about the thunderstorm structure and lightning in hurricanes.
- The first pilotless robotic aircraft (called an *aerosonde*) successfully flown across the Atlantic Ocean. In

the future, these planes may fly into hurricanes, providing timely and continuous crucial weather data at less cost than airplane reconnaissance.

- The largest coordinated measurements to date of landfalling hurricanes (Bonnie, Charlie, Earl, and Georges) by several groups of university scientists using portable instrumented towers, a portable wind profiler, and a portable Doppler radar.

Although these field programs are exciting, one is also reminded about how desperately such activities are needed. Two tragic 1998 hurricanes provide examples of why a better understanding of hurricanes is essential.

Hurricane Mitch (1998) produced an estimated 75 inches of rainfall in the mountainous regions of Central America, resulting in floods and mud slides that destroyed the entire infrastructure of Honduras and devastated parts of Nicaragua, Guatemala, Belize, and El Salvador. Whole villages and their inhabitants were swept away in the torrents of flood water and deep mud that came rushing down the mountainsides, destroying hundreds of thousands of homes and killing an estimated 9,000 to 18,000 people. Mitch was the most deadly hurricane to strike the Western Hemisphere in the previous 200 years. Not since the Great Hurricane of 1780, which killed 22,000 people in the eastern Caribbean, was there a more deadly hurricane. The president of Honduras, Carlos Flores Facusse, claims the storm destroyed fifty years of progress.

Hurricane Georges (1998) left a trail of destruction in the Caribbean region and across the southern U.S. Gulf coast. Estimates indicate more than 500 people were killed (or are still missing) in the Dominican Republic, with about 100,000 made homeless; 70 percent of the bridges in the Dominican Republic were damaged or destroyed, and 90 percent of all plantation crops were ruined. Damages in the Dominican Republic are estimated at $1 billion. In Puerto Rico, $2 billion in damage is estimated, with more than 33,000 homes destroyed and 50,000 more suffering major or minor damage. Georges destroyed 75 percent of the coffee crop in Puerto Rico and 95 percent of its plantation crops. The United States witnessed hundreds of homes damaged or destroyed in the Florida Keys and devastating floods associated with rain and the storm surge in eastern Mississippi, southern Alabama, and western Florida.

The hurricane challenge facing mankind in the twenty-first century involves several components. Although we cannot control hurricanes, we can encourage better preparation in foreign countries so that similar tragedies can be avoided in the future. New concerns have also emerged in the U.S. as coastal population growth and property development has exploded during this period. In fact, currently U.S. population increases are largest in coastal communities. The population has grown on average by 3–4 percent a year in hurricane-prone regions, and in some states like Florida the growth is greater. As a consequence, property damage costs due to hurricanes have skyrocketed in the 1990s, placing the insurance industry on the brink of financial chaos should a major hurricane hit a metropolitan area. In fact, many coastal residents are finding it difficult to obtain insurance now.

These congested coastal regions are also toying with human catastrophe since seashore development has proceeded without seemingly any considerations for speedy evacuation. Evacuation of many metropolitan areas now requires over two days. However, two-day hurricane track forecasts contain considerable error. Should a hurricane unexpectedly change course, there may not be enough time to evacuate a region, leading to a potential large loss of life.

As this book goes to press, another tragic reality of hurricanes becomes apparent—flash flooding. In North Carolina, emergency workers and Marines have rescued more than 1,500 people stranded as a result of the torrential rains produced by Hurricane Floyd. Dozens have drowned, and 30,000 homes in North Carolina are under water.

It is these financial and human consequences that motivate this publication. The purpose of this book is to provide background information on issues, people, organizations, and publications related to hurricanes and to provide guidance on where additional information may be obtained about a specific topic. Chapter 1 provides a broad introduction to hurricanes, ranging from hurricane formation to forecasting procedures to mitigation issues. Chapter 2 provides a chronology of forecast and scientific advances with regard to hurricanes. Chapter 2 also provides a descriptive timetable of significant U.S. landfalling hurricanes during the twentieth century, as well as a listing of hurricanes that have changed history. Chapter 3 contains biographies of important hurricane scientists and forecasters. Chapter

4 contains tabular data, interesting letters from survivors of hurricanes, and important documents. Several articles from the Federal Emergency Management Agency (FEMA) about hurricane preparedness and recovery issues are also included in Chapter 4. Chapter 5 describes relevant organizations that are involved in hurricane forecasting, research, and mitigation. Chapter 6 contains a comprehensive description of publications related to hurricanes. Chapter 7 lists electronic resources, such as videos and CD-ROMs containing hurricane footage and hurricane information, as well as useful Internet sites. The extensive appendixes include conversion tables, hurricane tracking information, and a thorough glossary.

Acknowledgments

This book could not have been completed without the assistance of many people. Dani Whitfield provided essential typing and clerical assistance. The following students assisted in this manuscript's preparation: Rafael Mahecha, Samir Mehta, Vasanthi Budamgunta, and Umesh Reddy. I'd like to thank the following institutions for sponsoring my visits while preparing this book: NOAA's Hurricane Research Division, the University of Maryland, and Colorado State University. I am indebted to Dr. Chris Landsea for allowing me to use his hurricane library collection, for providing an insightful critique of this manuscript, and for recommending me to ABC-CLIO as an author. Kristi Ward also provided valuable feedback. Thanks to all the scientists who furnished their vitae. The following people provided factual information that enhanced this book's contents: Julie Burba, Dr. Greg Holland, Natalie Blanchard, Dr. Jack Bevens, Dr. Steve Lyons, Vince Zegowitz, Yukio Takemura, Dr. John Knaff, Paul Simons, Sim Aberson, Richard Henning, John Williams, Derek West, Neal M. Dorst, Gary Padgett, Charlie Neumann, Dr. Paul Croft, Miles Lawrence, Chip Guard, and Frank Wells. Finally, thanks to my wife Amy, my daughters

Katie and Megan, and my parents for supporting me during this busy period of my life.

I gratefully acknowledge Dr. Bill Gray for opening many opportunities involving the study of hurricanes while I was a graduate student at Colorado State University.

This book was supported by the NASA Commercial Remote Sensing Program at Stennis Space Center through grants numbered NAS13-98033 and NAS13-564-SSC-127. The Mississippi Space Commerce Initiative (a partnership between the University of Mississippi, the Mississippi Department of Economic and Community Development, and Stennis Space Center) also sponsored this book through NASA grant NAG5-6209 and by the Office of Naval Research through grant number N00014-97-1-0811 under Bob Abbey. My interaction with User Systems Enterprises, Inc., at Stennis Space Center also positively contributed to this work.

An Overview of Hurricanes

1

A *hurricane,* which is a Caribbean Indian word for "evil spirit and big wind," is a large rotating system of oceanic tropical origin with sustained surface winds of at least 74 miles per hour (mph) somewhere in the storm. Due to the earth's rotation, these storms spin counterclockwise in the Northern Hemisphere and clockwise in the Southern Hemisphere; both types of hemispheric spins are referred to as *cyclonic rotation.* These storms occur worldwide and are called different names in different locations (Neumann 1993). In the Northwest Pacific across 180 degrees E, they are called *typhoons.* In the Philippines, they are called *chubasco.* Around Australia, they are called *severe tropical cyclones,*[1] whereas India calls them *severe cyclonic storms.* In deference to U.S. readers, storms in this 74-mph or faster category will be called hurricanes in this book.

A hurricane does not form instantaneously, but reaches this status in an incremental process. Initially such a tropical system begins as a *tropical disturbance* when a mass of organized, oceanic thunderstorms persists for 24 hours (NHC 1998). Sometimes partial rotation is observed, but this is not required for a system to be designated a tropical disturbance. The tropical disturbance becomes classified as a *tropical depression* when a closed circulation is first observed and sustained winds are less than 39 mph everywhere in the storm. When these sustained winds increase to 39 mph somewhere in the storm, it is then classified as a *tropical storm* and given a name.

It is important to note that these categories are defined by *sustained winds*, not instantaneous winds. Sustained winds are the average air speed over a period of time at roughly 33 feet above the ground. In the Atlantic and Northern Pacific Oceans, this averaging is performed over a 1-minute period, and in other ocean basins over a 10-minute period (Neumann 1993). The actual wind will be faster or slower than the sustained wind at any instantaneous point in time. Therefore, a hurricane with maximum sustained winds of 90 mph will actually contain gusts of 100 mph or more. Also, one should note that these categories are defined by *maximum winds* somewhere in the storm, almost always near the center, and that winds may be slower in other parts of the storm. For example, maximum sustained winds of 90 mph may be concentrated only in the northeast section near the hurricane center, with the southwest quadrant containing weaker winds.

By international agreement, the most general term for all large cyclonically rotating thunderstorm complexes over tropical oceans is *tropical cyclones.* The category tropical cyclones includes depressions, tropical storms, and hurricanes, in addition to other large, tropical, cyclonically rotating thunderstorm complexes that contain distinctly different temperature and organization characteristics. Examples include *monsoon depressions,* which bring moderate winds and heavy rain to India, and *subtropical cyclones,* which produce heavy rains and moderate winds in the Atlantic and Pacific Oceans. This book focuses on depressions, tropical storms, and hurricanes.[2] The details of the development process from tropical disturbance to hurricane will be discussed later, but the seasonal location of these storms will be examined first.

When and Where Do Hurricanes Occur?

Many U.S. residents perceive the North Atlantic Ocean basin as a prolific producer of hurricanes because of the publicity these storms generate. In reality, the North Atlantic is generally a marginal basin in terms of hurricane activity. Every tropical ocean except the South Atlantic and Southeast Pacific contains hurricanes; several of these tropical oceans produce more hurricanes annually than the North Atlantic. For example, the most active ocean basin in the world—the Northwest Pacific—averages 17 hurricanes per year. The second most active is the eastern North Pacific, which averages 10 hurricanes. In contrast, the North Atlantic mean annual number of hurricanes is 6. Table 1.1 summarizes each basin's mean number of hurricanes and total storms (Landsea 1999).

Hurricanes are generally a summer phenomenon, but the length of the hurricane season varies in each basin, as does the peak of activity (Table 1.2). For example, the Atlantic hurricane season officially starts on June 1 and ends November 30, but most

TABLE 1.1
Mean Number of Storms per Year

Tropical Ocean Basin	Mean Annual Storms (includes tropical storms and hurricanes)	Mean Annual Tropical Storms	Mean Annual Hurricanes	Mean Number of Hurricanes with Sustained Winds Greater than 100 mph (designated as major hurricanes)
Northwest Pacific (1970–1995)	27	10	17	8
Northeast Pacific (1970–1995)	17	7	10	5
Eastern Australia & Southwest Pacific (1969/70–1994/95)	10	5	5	2
Western Australia & Southeast Indian (1969/70–1994/95)	7	4	3	1
North Atlantic (1944–1995)	10	4	6	2
Southwest Indian (1969/70–1994/95)	10	5	5	2
North Indian (1970–1995)	5	3	2	Between 0 and 1
South Atlantic	0	0	0	0
Southeast Pacific	0	0	0	0
Global	86	38	48	20

Note: Mean number of hurricanes, total storms, tropical storms, hurricanes, and major hurricanes per year in all tropical ocean basins. Dates in parentheses provide the years for which accurate records were available through 1995. A tropical storm contains sustained surface wind speeds between 39 and 73 mph. A hurricane contains sustained surface winds greater than 73 mph. Hurricanes that contain winds faster than 110 mph are called major hurricanes.
Source: Adapted from Landsea, C. W., 1999. Climate Variability of Tropical Cyclones: Past, Present, and Future. In *Storms,* Routledge Press, United Kingdom, edited by Roger Pielke, Sr.

TABLE 1.2
Typical Hurricane Seasons

Tropical Ocean Basin	Hurricane Season	Active Regime	Peak Day(s)
Northwest Pacific (1945–1988)	Year Round	July 1–December 1	September 1
Northeast Pacific (1966–1989)	May 15–November 30	June 1–November 1	August 25
Eastern Australia and Southeast Pacific (1958–1988)	October 15–May 1	December 1–April 1	March 1
Western Australia and Southeast Indian (1958–1988)	October 15–May 1	January 1–April 1	January 15 and February 25
North Atlantic (1886–1989)	June 1–November 30	August 15–October 15	September 10
Southwest Indian (1947–1988)	October 15–May 15	December 1–April 15	January 15 and February 20
North Indian (1891–1989)	April 1–December 30	April 15–June 1 September 15– December 15	May 15 and November 10

Note: The length of the official hurricane season, when the season is most "active," and day when the season typically peaks for all tropical ocean basins. Active is subjectively defined relative to each basin's hurricane history (dates are shown in parentheses), and is defined as the period when most tropical storms and hurricanes occur.
Source: Adapted from Neumann, C. J., 1993. Global Overview. In *Global Guide to Tropical Cyclone Forecasting.* World Metereological Organization Technical Document, WMO/TD No. 560, Tropical Cyclone Programme, Report No. TCP-31, Geneva, Switzerland, Chapter 1. Edited by G. Holland.

tropical storms and hurricanes form between August 15 and October 15. The Atlantic hurricane season "peaks" around September 10. In contrast, hurricanes occur year-round in the western North Pacific, with a longer active regime lasting from July to November and peaking around September 1. In general, this pattern exists for other basins (with a late summer peak), although Southern Hemisphere hurricanes are six months out of phase. The exception is the North Indian basin, where there are two peaks, one in May and one in November (the reason for this will be discussed later).

How a Hurricane Forms

Hurricane formation occurs in two distinct phases. The first phase is called the *genesis stage* and includes tropical disturbances and tropical depressions. The second phase includes tropical storms and hurricanes and is called the *intensification stage.* These phases are separated because many disturbances and a few depressions never reach tropical storm intensity and eventually dissipate.

Genesis Stage

Tropical disturbances form in regions where surface air "piles up," which is known as *convergence*. Where convergence occurs, the atmosphere will respond by causing air to rise or sink. Since air experiencing convergence at ground level cannot penetrate a solid surface like water or land, it must ascend to balance this accumulation of air. As air rises, at higher levels in the atmosphere it will change from a gas to a liquid state and form the base of a cloud. Once the air has saturated to a liquid state, ascent is enhanced because in the warm, moist tropics the atmosphere is in a state of *instability*. In an unstable atmosphere, saturated air forced upward by convergence is less dense than surrounding unsaturated air and therefore accelerates upward, forming towering puffy clouds. The concepts of cloud formation and atmospheric instability are beyond the scope of this book, and the reader is referred to other meteorology books on the subject (Ahrens 1994; Nese and Grenci 1998; Danielson, Levin, and Abrams 1998).

Several conditions must simultaneously exist for a tropical disturbance to reach the tropical storm stage. First, the disturbance must be in a *trough*, defined as an elongated area of low pressure. (Pressure is the "weight" of the air above a given area of the earth's surface; its standard unit of measurement is the millibar, or mb). Furthermore, these troughs must contain a weak, partial cyclonic rotation; however, in general all troughs at least 10 degrees from the equator will obtain a partial cyclonic spin due to the earth's rotation. Troughs 10 degrees or further from the equator fall in three general categories: monsoon troughs, frontal troughs, and surface troughs.

Monsoon troughs occur in regions where air or water temperature actually increases away from the equator. In these regions, the *Intertropical Convergence Zone* (which is a region where Southern Hemisphere and Northern Hemisphere air converges near the equator) is displaced 10–20 degrees poleward, resulting in westerly winds on the equatorward side and easterly winds on the poleward side. A frontal trough is the remnant of a *front* (defined as the boundary between two air masses of different temperature and/or moisture characteristics) that has lost its contrasting temperature characteristic and entered the tropics. A surface trough encompasses all other kinds of troughs, such as those associated with a region of moisture contrast or broad thunderstorm complexes. Some surface troughs are even triggered by weather features 40,000 feet aloft.

The vast majority of genesis cases are associated with a monsoon trough. Many of the disturbances undergo the transition to tropical depression and then to tropical storm in the monsoon trough itself. However, a few experience this transition as *tropical waves*. A tropical wave is a trough, shaped like an upside down "V," resembling a wave, that has broken off from the monsoon trough and traveled westward a considerable distance before strengthening. In particular, most Atlantic hurricanes actually originate from tropical waves (more frequently called *easterly waves* in the United States) that broke off from the African monsoon trough and have propagated into the Atlantic! In the Atlantic about 55 to 75 tropical waves are observed each year, but only 10 to 25 percent of these develop into a tropical depression or beyond. Clearly, additional factors are required for genesis to occur.

The second condition required for genesis is a water temperature of at least 80°F. Heat transferred from the ocean to the air generates and sustains the thunderstorms in the disturbance through the instability mechanism discussed earlier. The third genesis condition is weak *vertical wind shear,* defined as the difference between wind speed and wind direction at 40,000 feet and the surface. In other words, for genesis to occur the wind must be roughly the same speed and blowing from the same direction at all height levels in the atmosphere.

Under these conditions, the embryo of a hurricane is born. The ascending air in the disturbance stimulates low-level inflow toward the center. This inflow slowly increases the cyclonic circulation of the disturbance, similar to the mechanism through which an ice skater spins faster as the skater pulls the arms inward (a process called conservation of angular momentum). When a closed circulation is observed, the disturbance is upgraded to a tropical depression. As long as the depression remains over warm water in a low vertical wind shear environment, the system will likely develop. Weak wind shear is a crucial factor because it allows vertical orientation of the thunderstorms and maintains the low-level inflow. Should the disturbance move into an environment where winds change dramatically with height, the thunderstorms are "torn apart" in different directions, the structure of the system is disrupted, and inflow weakens. Should adverse wind shear persist, or should the system move over colder water or over land, the disturbance or depression will weaken and eventually dissipate.

If optimum conditions persist, the sustained cyclonic winds will continue to slowly increase. Typically, the genesis time frame of disturbance and depression lasts for several days or longer. However, well-organized, rotating disturbance over warm water in weak wind shear can evolve much more quickly. When the cyclonic sustained winds increase to 39 mph somewhere in the disturbance, the depression is upgraded to a tropical storm. At this point, the intensification stage begins.

Intensification Stage

For a tropical storm to intensify into a hurricane, the same conditions that allowed its initial development (warm water and weak wind shear) must continue. Should this favorable environment persist, the rate of development increases compared to the genesis stage. This is because as the wind increases, more heat and moisture is transferred from the ocean to the air. The column of air begins to warm, which lowers surface pressure. Air will flow from higher pressure to lower pressure, trying to redistribute the atmosphere's weight, resulting in faster winds. In addition, the faster cyclonic winds also enhance convergence. Both factors increase thunderstorm production and low-level inflow. A feedback mechanism now occurs in which faster cyclonic winds breed more potent thunderstorms, which drop central surface pressure more and create stronger inflow, which in turn breeds faster cyclonic winds, and so on. Under favorable environmental conditions, a tropical storm can "spin-up" rather quickly, with winds increasing an additional 50 mph or more in a day. When sustained winds reach 74 mph somewhere in the storm, it is classified as a hurricane.

Water temperature is unquestionably linked to these storms' development. Hurricanes rarely form over water colder than 80F (although exceptions do occur, for reasons still not understood). They also weaken dramatically if a mature system moves over water colder than 80°F or if they make landfall, because their heat and moisture source has been removed. For a

hurricane to maintain thunderstorms through atmospheric instability, warm, moist surface air is required near the low-pressure
center. This warmth is provided by heat transfer from warm
ocean water, because otherwise air flowing toward lower pressure would expand and cool. (To convince yourself of this, let air
out of a tire and feel how cool it is. When one lets air out of a
high-pressure tire, the air expands as it enters the lower pressure
environment and cools. This cooling occurs because the motion
of gas molecules slows down as the air expands, and temperature
is essentially a measurement of molecular motion.) In other
words, heat flux from the warm water compensates for expansional cooling due to lower pressure, maintaining warm surface
air near the storm center.

When a hurricane moves over cold water, expansional
cooling dominates, which stabilizes the atmosphere; the thunderstorms disintegrate; and the hurricane weakens. When a
hurricane moves over land, the weakening occurs even faster
because not only has the surface heat flux been lost but so has
the moisture source for cloud formation. Land also has more
friction than water, which weakens landfalling hurricanes, but
this influence is minimal compared to the loss of heat and moisture flux.

Although warm water is significant, it is nearly as important that the warm water be at least 200 feet deep. Hurricanes
generate huge oceanic waves that mix the water to great depths.
If the warm water only covers a thin film at the top, the hurricane
will bring colder water to the surface and cut off the storm's
warm water energy supply, thereby weakening the system. In
fact, sometimes a hurricane will kill itself when it becomes stationary for an extended period of time. Should a hurricane stop
moving for several days, it can mix the ocean so much that all the
warm water is replaced by cold water, and the hurricane dissi-
pates. An example is Hurricane Roxanne in 1995 when it became
stationary in the Bay of Campeche.

The warmer the water, the greater are the chances for genesis, the faster is the rate of development, and the stronger these
storms can become. Under conditions of prolonged weak wind
shear and water temperature greater than 85°F, sustained winds
may reach almost 200 mph. Table 1.3 shows the maximum potential intensity a tropical storm or hurricane can achieve for a given
water temperature. Fortunately, few hurricanes over warm water

TABLE 1.3
Maximum Potential Hurricane Intensity for a Given Water Temperature

Water Temperatures (°F)	Maximum Potential Sustained Wind for a Hurricane (in mph)
70	101
71	104
72	107
73	110
74	114
75	118
76	122
77	127
78	132
79	138
80	145
81	152
82	160
83	169
84	178
85	189
86	201

Note: Maximum potential intensity (as measured by sustained winds in mph) of a mature hurricane for a given water temperature (in °F). Please note that this table is not valid for the genesis stage because hurricanes do not form over water colder than 80° F. However, it is valid for a mature hurricane moving over colder water.
Source: Adapted from DeMaria, M., and J. Kaplan, 1994. Sea Surface Temperature and the Maximum Intensity of Atlantic Cyclones. *J. Climate,* 7:1324–1334.

reach their potential because some inhibiting factor (such as wind shear, landfall, or movement over colder water) occurs.

As shown earlier, hurricanes are most numerous and strongest in late summer for most ocean basins. This is because the three favorable conditions—warm water, weak wind shear, and cyclonic disturbances—are optimum in late summer. In particular, water's temperature peaks in late summer, which seems paradoxical because the longest day is in June. However, the days are still longer than nights until fall, therefore the water is still accumulating heat into late summer. The monsoon troughs are most active in late summer as well, and large-scale circulation patterns favor weak wind shear in late summer. The exception is the North Indian oceans, where the season peaks in spring and fall (Table 1.2). Two factors create this double peak: Strong wind shear occurs over India during the summer, and the Indian monsoon moves inland during the summer.

Naming Hurricanes

When a tropical depression is upgraded to a tropical storm, it is assigned a name. Before this practice was started, tropical storms and hurricanes were identified in many confounding ways. Some legendary storms have been inconsistently named for the holiday they occurred on (Labor Day Hurricane of 1935), the nearest saint's day (Hurricane Santa Anna of 1825), the area of landfall (Galveston Hurricane of 1900), or even for a ship (Racer's storm in 1837). However, most storms before the 1940s never received any kind of designation.

During forecast operations, the situation was just as perplexing. At first forecasters used cumbersome latitude-longitude identifications. This naming convention was too long and was confusing when more than one storm was present in the same ocean basin. Right before World War II, this procedure was changed to a letter designation (e.g., A-1943).

The naming of hurricanes was amusingly initiated by the Australian forecaster Clement Wragge, who occasionally named severe storms after politicians he was displeased with. Wragge is apparently also the first person to give hurricanes female names. However, the naming convention began in earnest when western North Pacific World War II forecasters began informally naming tropical storms after their girlfriends and wives. This practice became entrenched in the system, and beginning in 1945 Northwest Pacific storms officially were given female names.

Atlantic storms were officially given names starting in 1950 that were radio code words (e.g., Able, Baker, Charlie). In 1953 the U.S. Weather Bureau switched to a female list of names. The Northeast Pacific began this convention for the Hawaii region in 1959 and for the rest of the basin in 1960. Southwest India storms were first named during the 1960–1961 season. Australian storms were first named in 1964. Storms are not named in the North Indian oceans. Male names were added to the lists in the following years: Northwest Pacific in 1979, Northeast Pacific in 1978, Atlantic in 1979, Australia in the 1974–1975 season, and Southwest India in the 1974–1975 season.

Table 1.4 shows six lists of names for the Atlantic Ocean. Each list is used alphabetically for a particular year, then is recycled six years later. Should an Atlantic storm be very destructive or should it have a noteworthy impact on human lives or the economy, its name is retired (Table 1.5) and replaced by a new name beginning with the same letter (LePore 1996). The letters "Q," "U," "X," "Y," and "Z" are not used by the National Hurricane Center (NHC), leaving 21 names on the list. Should more than 21 Atlantic storms occur in a year (which has never been recorded), the next storms would be given Greek letter names (Alpha, Beta, etc.). Other oceans contain a similar list of names. All names are determined by the World Meteorological Organization (WMO).

TABLE 1.4
Tropical Storm and Hurricane Names for the Atlantic Ocean

1999	2000	2001	2002	2003	2004
Arlene	Alberto	Allison	Arthur	Ana	Alex
Bret	Beryl	Barry	Bertha	Bill	Bonnie
Cindy	Chris	Chantal	Cristobal	Claudette	Charley
Dennis	Debby	Dean	Dolly	Danny	Danielle
Emily	Ernesto	Erin	Edouard	Erika	Earl
Floyd	Florence	Felix	Fay	Fabian	Frances
Gert	Gordon	Gabrielle	Gustav	Grace	Gaston
Harvey	Helene	Humberto	Hanna	Henri	Hermine
Irene	Isaac	Iris	Isidore	Isabel	Ivan
Jose	Joyce	Jerry	Josephine	Juan	Jeanne
Katrina	Keith	Karen	Kyle	Kate	Karl
Lenny	Leslie	Lorenzo	Lili	Larry	Lisa
Maria	Michael	Michelle	Marco	Mindy	Matthew
Nate	Nadine	Noel	Nana	Nicholas	Nicole
Ophelia	Oscar	Olga	Omar	Odette	Otto
Philippe	Patty	Pablo	Paloma	Peter	Paula
Rita	Rafael	Rebekah	Rene	Rose	Richard
Stan	Sandy	Sebastien	Sally	Sam	Shary
Tammy	Tony	Tanya	Teddy	Teresa	Tomas
Vince	Valerie	Van	Vicky	Victor	Virginie
Wilma	William	Wendy	Wilfred	Wanda	Walter

Note: Unless a hurricane name is retired and replaced by a new name, the list is recycled every six years (i.e., the names used in 1999 will be used again in 2005, the names used in 2000 will be used again in 2006, etc.).
Source: National Hurricane Center homepage. http//www.nhc.noaa.gov.

TABLE 1.5
Retired Atlantic Hurricane Names and the Year of Their Occurrence

Agnes 1972, Alicia 1983, Allen 1980, Andrew 1992, Anita 1977, Audrey 1957
Betsy 1965, Beulah 1967, Bob 1991
Camille 1969, Carla 1961, Carmen 1974, Carol 1965, Celia 1970, Cesar 1996, Cleo 1964, Connie 1955
David 1979, Diana 1990, Diane 1955, Donna 1960, Dora 1964
Edna 1968, Elena 1985, Eloise 1975
Fifi 1974, Flora 1963, Fran 1996, Frederic 1979
Gilbert 1988, Gloria 1985, Gracie 1959, Georges 1998
Hattie 1961, Hazel 1954, Hilda 1964, Hortense 1996, Hugo 1989
Inez 1966, Ione 1955
Janet 1955, Joan 1988
Klaus 1990
Luis 1995
Marilyn 1995, Mitch 1998
Opal 1995
Roxanne 1995

Source: Landsea, C., 1998. Frequently Asked Questions: Hurricanes, Typhoons, and Tropical Cyclones. http://www.aoml.noaa.gov/hrd/tcfaq/.

Hurricane Structure

The structure of a hurricane is certainly one of the most fascinating, awesome, and bizarre features in meteorology. Distinct cloud patterns exist for each stage of a hurricane's life cycle. These patterns are so unique that a meteorologist can estimate the intensity of a depression, tropical storm, or hurricane based solely on cloud organization and cloud height using a methodology known as the *Dvorak technique* (Dvorak 1975). During genesis, a mass of clouds with a weak rotation is typically first observed, known as "stage 1 of genesis." Usually these clouds will temporarily dissipate, leaving a residual circulation, although in the case of a tropical wave a cloud pattern resembling an upside down "V" is sometimes observed. When the clouds return and the circulation increases, "stage 2 of genesis" commences, which coincides with the depression stage.

Clouds start forming a curved pattern, and when the winds reach tropical storm strength, intricate patterns emerge. In the center, where convergence is strongest, a 100-mile-wide region of clouds grows to around 50,000 feet in height, surrounded by less tall but still potent thunderstorm bands out to 500 miles from the center. These curved thunderstorm bands, known as *spiral bands,* are clustered about 20 miles apart, with light to moderate rain between them. In the periphery of a tropical storm, wind speed will fluctuate, with fastest sustained winds of 30–40 mph in the spiral bands. As one approaches the center of a tropical storm, winds will consistently increase, with the strongest winds close to the center.

As the winds in the center increase to hurricane strength (74 mph), a ring of thunderstorms form around the center, known as the *eyewall*. In this eyewall are the fiercest winds and the most dangerous part of a hurricane. In the center itself, a clear region devoid of clouds forms, known as the *eye*. In the eye, winds become weak, even calm! This transition from hurricane force winds to calm is rather sudden (often within minutes) and is truly one of the most bizarre features of any weather system.

The eye size can vary from 10 miles wide to 50 miles wide (there have even been cases of 100-mile-wide eyes in the western North Pacific Ocean). Typically an eye starts at about 25 miles wide during the transition from tropical storm to hurricane. As the hurricane intensifies, usually the eye contracts. However, sometimes a second eyewall forms outside the original eyewall, known as the *concentric eyewall cycle* (Willoughby et al. 1982). This outer eyewall cuts off the inflow to the inner eyewall, causing the inner one to weaken and dissipate. As the eye expands, temporary weakening occurs. The outer eyewall will begin to contract inward to replace the inner eyewall, and about 12 to 24 hours later, intensification resumes.

The cause of eye formation is still not understood, but meteorologists generally agree that near 74 mph the strong rotation impedes inflow to the center, causing air to instead ascend at 10 to 20 miles from the center. Consider what happens as one drives at fast speeds in a vehicle. As the driver makes a sharp turn, an "invisible" force makes the driver lean outward. This outward-directed force, called the *centrifugal force,* occurs because the driver's momentum wants to remain in a straight line, and since the car is turning, there is a tugging sensation outward. The sharper the curvature and/or the faster the rotation, the stronger is the centrifugal force.

As rotation increases in a developing hurricane, air is subjected to these outward accelerations, which counteracts inflowing air. The centrifugal force eventually dominates near hurricane strength, causing inflowing air to not reach the center and to instead ascend about 15 miles from the center. This strong rotation also creates a vacuum of air at the center, causing some of the air flowing out the top of the eyewall to turn inward and sink to balance this loss of mass. This subsidence suppresses cloud formation, creating a pocket of generally clear air in the center (although low-level short clouds or high-level overcast skies may still exist). People experiencing an eye passage at night often see stars. Trapped birds are sometimes seen circling in the eye, and ships trapped in a hurricane report hundreds of exhausted birds resting on their decks. The landfall of Hurricane Gloria (1995) on southern New England was accompanied by thousands of birds in the eye.

The sudden change of violent winds to a calm state is a dangerous situation for people ignorant about a hurricane's structure. Those experiencing the calm of an eye may think the hurricane has passed, when in fact the storm is only half over, with dangerous eyewall winds returning from the opposite direction within 20 minutes or less. Marjory Stoneman Douglas, the Florida Everglades activist, writes in the well-known book *Everglades: River of Glass* of the eye passage experience of survivors of the 1926 Miami hurricane:

> Late that night, in absolute darkness, it hit, with the far shrieking scream, the queer rumbling of a vast and irresistible freight train. The wind instruments blew away at a hundred twenty-five miles [per hour]. The leaves went, branches, the bark off the trees. In the slashing assault people found their roofs had blown off, unheard in the tumult. The water of the bay was lifted and blown inland, in streaming sheets of salt, with boats . . ., coconuts, debris of all sorts, up on the highest ridge of the mainland. . . .
>
> At eight o'clock next morning the gray light lifted. The roaring stopped. There was no wind. Blue sky stood overhead. People opened their doors and ran, still a little dazed, into the ruined streets. . . . Only a few remembered or had ever heard that in the center of a spinning hurricane there is that bright deathly stillness.
>
> It passed. The light darkened. The high shrieking came from the other direction as the opposite whirling thickness of the cyclonic cone moved on over the darkened city.

In fact, many killed in the Labor Day Hurricane had wandered outside during the passage of the eye. In addition to the return of fierce winds, sometimes the highest storm surge (to be discussed later) occurs on the backside of the eye.

Outside the eyewall region, the weaker spiral bands accompanying the hurricane typically affect a large area. On average, the width of a hurricane's cloud shield is about 500 miles, but it may vary tremendously. For example, the cloud shields of Hurricane Gilbert (1988) and Hurricane Allen (1980) covered an impressive one-third of the Gulf of Mexico. Some are giants, such as Supertyphoon Tip (1979), which was twice the size of Gilbert or Allen! In contrast, some hurricanes are very small (typically called *midgets* by meteorologists), such as the Labor Day Hurricane of 1935 and Tropical Cyclone Tracy, which hit Australia in 1974. However, storm size does not correlate with storm intensity. Both midgets and giant hurricanes can have sustained eyewall winds in excess of 100 mph. In fact, the Labor Day Hurricane and Tropical Cyclone Tracy were extremely destructive.

Hurricane Destruction

Coastal communities devastated by strong hurricanes usually take years to recover. Many forces of nature contribute to the destruction. Obviously, hurricane winds are a source of structural damage. As winds increase, pressure against objects increases at a disproportionate rate. For example, a 25-mph wind causes about 2.3 pounds of pressure per square foot. In 75-mph winds, that force becomes 17 pounds per square foot. When the wind exceeds a building's design specifications, structural failure occurs. Debris is also propelled by the winds, compounding the damage.

Isolated pockets of enhanced winds occur in hurricanes, too. As hurricanes make landfall, interactions with the thunderstorms form columns of rapidly rotating air in contact with the ground, known as *tornadoes*. Also accompanying the thunderstorms are *downbursts*, or areas where heavy rainfall accelerates air to the ground and spreads out at speeds greater than 100 mph. In addition, another phenomena was documented in Hurricane Hugo (1989) and Hurricane Andrew (1992) called *mesoscale vortices* (Willoughby and Black 1996; Fujita 1993). These are whirling vortices 150 to 500 feet wide that form at the boundary of the eyewall and eye where there is a tremendous change in wind speed. The updrafts in the eyewall stretch the vortices vertically, making them spin faster, with winds up to 200 mph. Damage by these three wind phenomena occurs in narrow swaths inland, although sometimes it is difficult to discern which wind event caused a particular swath's destruction during the poststorm analysis.

Floods produced by the rainfall can also be quite destructive, and they are currently the leading cause of hurricane-related fatalities in the United States. Hurricanes average 6 to 12 inches of rain at landfall regions, but this amount varies tremendously and is not necessarily proportional to hurricane intensity. In fact, weaker tropical storms often produce greater amounts of rain. For example, Tropical Storm Claudette (1979) brought a U.S. record of 42 inches of rain in 24 hours near Houston, Texas, causing $400 million (in 1979 dollars) of damage, making it the costliest Atlantic tropical system that was not a hurricane. Rainfall accumulation generally increases for slow-moving storms. Hurricane Danny (1997) sat over Mobile Bay for almost one day before moving inland, dumping at least 37 inches of rain on coastal Alabama in 36 hours, of which 26 inches fell in 7 hours!

Heavy rainfall is not just confined to the coast. The remnants of hurricanes can bring heavy rain far inland, which is particularly dangerous in hills and mountains where acute concentrations of rain turn tranquil streams into raging rivers in a matter of minutes. In addition, mountains "lift" air in hurricanes, increasing cloud formation and rainfall. Rainfall rates of one to two feet per day are not uncommon in mountainous regions when hurricanes pass through. In fact, the highest hurricane rainfall amounts have occurred in the mountains of La Reunion Island (see Chapter 4 for details).

Some examples regarding rainfall damage and fatalities follow: Hurricane Camille (1969), which made landfall in Mississippi, dumped 30 inches of rain in 6 hours in the Blue Ridge Mountains, triggering flash floods and mudslides that killed 114 people in Virginia and 2 in West Virginia. One of the most widespread floods in U.S. history was caused by Hurricane Agnes (1972), which caused 188 deaths and $2.1 billion (in 1972 dollars) in property damage along most of the eastern United States. Mud slides are also often a problem in hilly terrain, burying homes and destroying property, particularly in underdeveloped mountainous countries where warning systems, hurricane preparedness, and infrastructure are insufficient. Many of these countries have high terrain near the coast, where mud slides and floods can occur with tragic results. Eastern Hemisphere countries sustain the worst fatalities because more storms occur there. A few of many examples include the Philippines (Typhoon Kelly in 1981 with 140 killed and Tropical Storm Thelma in 1991 with 3,000 killed), China (Typhoon Peggy in 1986 with 170 killed, Typhoon Herb in 1996 with 779 killed, and Typhoon Nina in 1975 with at least 10,000 killed), and Korea (Typhoon Thelma in 1987 with 368 killed or missing). However, Western Hemisphere countries also suffer sizable casualties from flooding and mudslides, such as in the Caribbean nations (Hurricane Fifi in 1974 with 8,000 to 10,000 killed, Hurricane Flora in 1963 with 8,000 killed, and Hurricane Gordon in 1994 with 1,145 killed), Mexico (Hurricane Gilbert in 1988 with 202 killed and Hurricane Pauline in 1997 with 230 killed), and Central America (Hurricane Mitch in 1998).

As discussed in the preface, the recent Hurricane Mitch (1998) produced an estimated 75 inches of rainfall in the mountainous regions of Central America, resulting in floods and mud slides that destroyed the entire infrastructure of Honduras and devastated parts of Nicaragua, Guatemala, Belize, and El Salvador. Whole villages and their inhabitants were swept away in the torrents of flood water and deep mud that came rushing down the mountainsides, destroying hundreds of thousands of homes, and killing more than 9,000 people with an estimated 9,000 missing and presumed dead. Mitch was the most deadly hurricane to strike the Western Hemisphere in 200 years, and the president of Honduras, Carlos Flores Facusse, has claimed the storm destroyed 50 years of progress. Typically in all the flood examples listed above, the number of homeless was in the 10,000 to over 100,000 range.

Although all these elements (wind, rain, floods, and mud slides) are obviously dangerous, historically most people have been killed in the *storm surge,* defined as an abnormal rise of the sea along the shore. The storm surge, which can reach heights of 20 feet or more, is caused by the winds pushing water toward the coast. As the transported water reaches shallow coastlines, bottom friction slows their motion, causing water to pile up. Ocean waters begin to rise gradually, then quite quickly as the storm makes landfall (it does *not* occur as a tidal wave, as depicted in at least one Hollywood movie!).

Some factors that determine a storm surge's height include storm intensity, storm size, storm speed, and the angle at which the hurricane makes landfall. The storm surge increases with storm intensity, size, and speed. The storm surge is also greater when landfall is perpendicular to the coastline, since some of the surge will be deflected offshore when storms land at a sideways angle. The shape of the coastal estuary is another important component, because coastal points and channels tend to enhance the surge in certain regions. One more element is the proximity of shallow water near the coast. Low-lying regions adjacent to shallow seas (such as the Gulf of Mexico in the southern United States and the Bay of Bengal bordering Bangladesh and India) are particularly vulnerable to the storm surge, because more water piles up before inundating the coast. Another minor contribution to the storm surge is the low pressure of a hurricane, which allows water to expand a little (known as the *inverted barometer effect*). For every 10-mb pressure drop, water expands 3.9 inches. The storm surge is always highest on the side of the eye corresponding to onshore winds, which is usually the right side of the point of landfall in the Northern Hemisphere, called the *right front quadrant.* Winds are also fastest in the right front quadrant because storm motion (which averages about 10 mph but varies substantially) is added to the hurricane's winds.

The total elevated water includes two additional components—the astronomical tide and ocean waves. The astronomical tide results from gravitational interactions between the earth, moon, and sun, generally producing two high and two low oceanic tides per day. Should the storm surge coincide with the high astronomical tide, additional feet will be added to the water level, especially when the sun and moon are aligned, which produces the highest oceanic tides (known as *syzygy*). The total water elevation caused by the storm surge, astronomical tides, and wave setup is known as the *storm tide.* Therefore, the storm surge is officially defined as the difference between the actual water level under the hurricane's influence and the level due to astronomical tides and wave setup. In practice, water level observations during posthurricane surveys are always storm tides, and it is difficult to distinguish between storm surge and storm tide water elevations. Therefore, the two terms are used interchangeably.

Water is very powerful, weighing some 1,700 pounds per cubic yard, and therefore most inundated structures pounded by waves and the storm surge will be demolished. Ocean currents set up by the surge, combined with the waves, can severely erode beaches, islands, and highways. Most people caught in a storm surge will be killed by injuries sustained during structural collapse or by drowning. Death tolls for unevacuated coastal regions can be terrible. The worst natural disaster in U.S. history occurred in 1900 when a hurricane-related storm surge measuring 8 to 15 feet inundated the island city of Galveston, Texas, and claimed over 6,000 lives. In 1893, nearly 2,000 were killed in Louisiana and 1,000 in South Carolina by two separate hurricanes.

Hurricane Camille (1969), with sustained winds of at least 180 mph, produced a storm surge of 23 feet in Pass Christian, Mississippi. As the storm surge penetrated far inland, Camille killed 137 people in Mississippi and 9 in Louisiana, including 20 of 23 people who ignored evacuation warnings and stayed for a "hurricane party." A survivor of the hurricane party, who clung to floating debris, traveled 10 miles or more in the storm surge. In Louisiana, Camille produced a storm surge that pushed water over both levees near the mouth of the Mississippi River, "removing almost all traces of civilization" as one U.S. Department of Commerce (1969) report states. Floods from the storm surge penetrated 8 miles inland in the Waveland-Bay St. Louis region, and in river estuaries the flood extended 20 to 30 miles upstream (Corps of Engineers 1970). When combined with the 116 Virginia flood deaths mentioned earlier, Camille killed a total of 262 people (Corps of Engineers 1970). The hurricane caused a total of $1.5 billion in property damage (in 1969 dollars), with total devastation on the immediate coastline and severe damage further inland.

However, these statistics pale compared to the lives taken in coastal India and surrounding countries by storm surges. The most vulnerable area is coastal Bangladesh, a huge river delta fertile enough to support large numbers of farmers and fishermen (about 1,500 people per square mile). Many are essentially nomads, staking claims on frequently changing temporary islands. The geography of this area favors large storm surges because the narrow inlet of the bay funnels large amounts of water inland and because shallow water extends 60 miles offshore. Storm surge warnings are issued more than a day in advance, but communicating this information is difficult because of the rural and nomadic nature of the population. Furthermore, many choose to ignore the warnings or find it too difficult to evacuate because no transportation infrastructure exists (Rosenfeld 1997). As a result, the worst storm surge fatalities in history have occurred in this area.

Six hurricanes hit Bangladesh between 1960 and 1997, killing at least 10,000 each time (Rosenfeld 1997). The most tragic incident occurred on November 12, 1970, when a hurricane with 125-mph winds caused a 20-foot storm surge that killed 300,000 people. This storm triggered a revolution against Pakistan that brought about Bangladesh independence. Unfortunately, this tragedy was still repeated in 1991 when 139,000 people perished in another Bangladesh hurricane. In general, this area has a history of hurricane-induced fatalities: 300,000 people were killed near Calcutta, India, in 1737; 20,000 were killed in Coringa, India, in 1881; and 200,000 were killed in Chittagong, Bangladesh, in 1876. Unfortunately, evacuation procedures and public response has not changed much since 1991, so the chance of similar tragedies remains high. China, Thailand, and the Philippines have also seen losses in the tens to hundred thousands in recent years due to storm surges.

The combination of heavy rainfall and a storm surge can be particularly devastating, especially along coastal streams. Under these conditions, residents will first experience the surge, which typically lasts for one day, followed by flooding from run-off that persists for weeks. For example, coastal homes in Pasagoula, Mississippi, were devastated by slow-moving Hurricane Georges (1998) because the region was inundated first by the storm surge and then by flooding from the Pasagoula River.

To quantify the expected levels of damage for a given hurricane intensity, Herbert Saffir and Robert Simpson devised the *Saffir-Simpson scale* (Simpson 1974).[3] This scale, only valid for the Atlantic Ocean, classifies hurricanes into five categories according to central pressure, maximum sustained winds, storm surge, and expected damage (Table 1.6). Although all categories are dangerous, Categories 3, 4, and 5 are considered *major hurricanes,* with the potential for widespread devastation and loss of life. Whereas only 21 percent of U.S. landfalling tropical systems are major hurricanes, they historically account for 83 percent of the damage in the United States (Pielke and Landsea 1998). On average, the Atlantic has two major hurricanes per year (Table 1.1). Fortunately, Category 5 hurricanes are infrequent in the Atlantic Ocean and seldom sustain themselves at such intensities for very long before weakening to a lower category. Only two Category 5 hurricanes have made landfall in the United States in the twentieth century (Hurricane Camille in 1969 and the Florida Keys' Labor Day Hurricane of 1935). Major hurricanes are much more common in the eastern North Pacific and in the western North Pacific. In fact, in the western North Pacific Ocean major hurri-

TABLE 1.6
The Saffir-Simpson Scale for Atlantic Hurricanes

Category	Central Pressure (approximate)		Maximum Sustained Winds in Mph	Storm Surge in Feet (approximate)	Potential Damage Scale	Damage
	mb	inches				
1 Minimal	> 979	> 28.91	74–95	4–5	1	Damage primarily to shrubbery, trees, foliage, and unanchored mobile homes. No real damage to building structures. Low-lying coastal roads inundated, minor pier damage, and some small craft in exposed anchorages torn from moorings.
2 Moderate	965–979	28.50–28.91	96–110	6–8	10	Considerable damage to shrubbery and tree foliage, some trees blown down and major damage to exposed mobile homes. Some damage to roofing, windows, and doors of buildings. Coastal road and low-lying escape routes inland cut by rising water two to four hours before arrival of hurricane center. Considerable pier damage, marinas flooded, and small craft torn from moorings. Evacuation of shoreline residences and low-lying island areas required.
3 Extensive	945–964	27.91–28.47	111–130	9–12	50	Large trees blown down. Foliage removed from trees. Structural damage to small buildings; mobile homes destroyed. Serious flooding at coast and many smaller coastal structures destroyed. Larger coastal structures damaged by battering waves and floating debris. Low-lying inland escape routes cut by rising waters between three and five hours before arrival of hurricane center. Low-lying inland areas flooded eight miles or more. Evacuation of low-lying structures within several blocks of shoreline possibly required.
4 Extreme	920–944	27.17–27.88	131–155	13–18	250	All signs blown down. Extensive damage to roofing, windows, and doors. Complete failure of roofs on smaller buildings. Flat terrain

(continues)

TABLE 1.6
(Continued)

Category	Central Pressure (approximate)		Maximum Sustained Winds in Mph	Storm Surge in Feet (approximate)	Potential Damage Scale	Damage
	mb	inches				
5 Catastrophic	<920	<27.17	>155	>18	500	10 feet or fewer above sea level flooded as far as six miles inland. Major damage to lower floors of coastal buildings from flooding, battering waves, and floating debris. Major erosion of beaches. Massive evacuation: all residences within 500 yards of shore and single-story residences on low ground within two miles of shore.
						Severe and extensive damage to residences and buildings. Small buildings overturned or blown away. Severe damage to windows and doors; complete roof failures on homes and industrial buildings. Major damage to lower floors of all structures less than 15 feet above sea level. Flooding inland as far as 10 miles. Inland escape routes cut three to five hours before arrival of storm center. Massive evacuation of residential areas on low ground within 5–10 miles of shore.

Note: In practice, the maximum wind speed determines the category. Many factors affect central pressure and storm surge, so these values are only estimates for a particular category. In fact, the storm surge may vary by a factor of two depending on the coastline's proximity to deep or shallow water. This scale is not valid for other ocean basins, because some countries use different definitions of sustained winds (i.e., Australia and India), contain different types of foliage with different damage thresholds than U.S. foliage, have different building construction standards (some of which may be further weakened by termites), and have coral reefs that modify storm surge damage. "Potential Damage Scale" provides a scale relative to a Category 1 hurricane, where a Category 1 hurricane is scaled as "1" (Pielke and Landsea 1998). For example, a Category 3 hurricane typically causes 50 times as much damage as a Category 1 hurricane.

Source: Simpson, R. H., 1974. The Hurricane Disaster Potential Scale. *Weatherwise*, 27:169, 186.

canes (or typhoons, as they are called there) are so common that a new term has been created for storms with sustained winds greater than 149 mph—*supertyphoons.* Some Australians give major hurricanes the colorful name "cock-eyed Bob."

Although hurricanes have wrought much misery, they can also be beneficial. Hurricanes often provide much-needed rain to drought-stricken coastlines. Their ocean interactions can flush bays of pollutants, restoring the ecosystem's vitality. After the record rainfall from Hurricane Claudette (1979) in Texas, fish were being caught in the northern industrialized reaches of Galveston Bay that had vanished for several years. Finally, in cruel Darwinian fashion, weak sea life and plants perish during a hurricane, leaving only the strong to survive and reproduce.

In this vein, sometimes hurricanes "correct" humanity's mistakes. For example, in the early 1900s nonnative foliage, such as Australian pine trees, had been planted on the tip of Key Biscayne, Florida (now the Bill Baggs State Park). These nonnative plants had few natural enemies in their new environment, and they quickly dominated plant life, resulting in a loss of natural habitat. However, these Australian nonnatives lacked the ability to withstand hurricane-force winds, and Hurricane Andrew destroyed them all in 1992. Park officials seized the opportunity to replant the park with native foliage.

Forecasting Hurricanes

Hurricanes are one of the most difficult phenomena to forecast in meteorology. A forecaster must understand all facets of meteorology for hurricane prediction because these storms encompass all weather processes, from individual thunderstorms to rainband physics, to air-sea coupling, to interaction of the hurricane itself with the surrounding atmosphere. These large- and small-scale weather features also interact with each other in complicated ways that even a computer can only crudely simulate and predict. Even worse, because hurricanes occur over the data-sparse ocean, forecasters have few observations to see the current state of a hurricane or to input into the computers. Without knowing what the storm is doing in the present, how can one anticipate its future? Forecasting hurricanes is indeed a challenge.

Government hurricane forecast centers exist worldwide (Table 1.7). The center charged with Atlantic and Northeast Pa-

TABLE 1.7

Hurricane Forecast Centers

Hurricane Forecast Centers	Regions of Responsibility
National Hurricane Center, Miami, Florida	Atlantic Ocean, Caribbean Sea, Gulf of Mexico, Eastern Pacific Ocean
Central Pacific Hurricane Center, Honolulu, Hawaii	Central Pacific Ocean
Naval Pacific Meteorology and Oceanography Center/ Joint Typhoon Warning Center, Pearl Harbor, Hawaii	All oceans in the Eastern Hemisphere and East Pacific Ocean
Regional Specialized Meteorological Center Tokyo Typhoon Center, Tokyo, Japan	Northwest Pacific Ocean
Royal Observatory, Kowloon, Hong Kong	Northwest Pacific Ocean
Bangkok Tropical Cyclone Warning Center, Thailand	Northwest Pacific Ocean
Fiji Tropical Cyclone Warning Center, Nadi Airport, Fiji	Australian/Southwest Pacific Ocean
New Zealand Meteorological Service, Willington, New Zealand	Australian/Southwest Pacific Ocean
Port Moresby Tropical Cyclone Warning Center, Boroco, NCD, Papua New Guinea	Australian/Southwest Pacific Ocean
Brisbane Tropical Cyclone Warning Center, Australia	Southeast Indian/Australian Ocean
Darwin Tropical Cyclone Warning Center, Australia	Southeast Indian/Australian Ocean
Perth Tropical Cyclone Warning Center, Australia	Southeast Indian/Australian Ocean
Regional Tropical Cyclone Advisory Center, Reunion Island	Southwest Indian Ocean
Sub-Regional Tropical Cyclone Warning Center, Mauritius Island	Southwest Indian Ocean
Sub-Regional Tropical Cyclone Warning Center, Madagascar	Southwest Indian Ocean
Philippine Atmospheric, Geophysical and Astronomical Services Administration (PAGASA), Quezon City, Philippines	Northwest Pacific Ocean
National Meteorological Center of the China Meteorological Administration (CSA); there are forecast centers in Dalian, Guanzhou, and Shanghai, China	Northwest Pacific Ocean
Korean Meteorological Administration (KMA), Seoul, Korea	Northwest Pacific Ocean
Hydro-Meteorological Service (HMS), Hanoi, Vietnam	Northwest Pacific Ocean
Bangladesh Meteorological Department, Dhaka, Bangladesh	North Indian Ocean
India Meteorological Department; there are forecast centers in Calcutta and Bombay	North Indian Ocean
Department of Meteorology and Hydrology, Yangon, Myanmar	North Indian Ocean
Department of Meteorology, Male, Maldives	North Indian Ocean

Source: Landsea, C. W., 1998. Frequently Asked Questions: Hurricanes, Tyhpoons, and Tropical Cyclones. http://www. aoml.noaa.gov/hrd/tcfaq/; Takemura, Y., 1998. Personal communication.

cific hurricane forecasts is the National Hurricane Center (NHC) located on the campus of Florida International University in Miami. All the hurricane centers use similar forecast procedures, so this book will discuss NHC operations.

For evacuation and emergency preparedness purposes, forecasters are most concerned with predicting where the storm will go and, should the hurricane threaten land, where the eye will make landfall. Of nearly equal importance is forecasting hurricane intensity and the width of the tropical and hurricane force winds around the hurricane. Forecast statements are also issued for expected rainfall and storm surge. A *tornado watch,* which means conditions are favorable for tornado development, is always issued preceding hurricane landfall by the nearest National Weather Service office. When a tornado has been observed by trained people called "weather spotters" or inferred by an instrument called Doppler radar (which can detect areas of rapid rotation in a thunderstorm), a *tornado warning* is issued by the National Weather Service.

Hurricanes have a reputation for being unpredictable at times. Hurricanes can suddenly turn, speed up, slow down, stall, or loop. They can also reform their center of rotation when thunderstorms are not uniformly distributed around the center, making the storm suddenly "jump" from one spot to the next (an example is Hurricane Earl in 1998). Track forecasts have improved considerably the past 20 years, but errors still occur. Intensity forecasts, however, currently exhibit little reliability and have not improved in the last 20 years. The process of anticipating whether a hurricane will strengthen, weaken, or not change intensity—and predicting how quickly any intensity change may occur—is still an unsolved forecast problem. Such uncertainties are fraught with potential disasters, as in the "near-miss" for Hurricane Opal (1995), described in the following section.

Hurricane Opal—A Close Call

Hurricane Opal formed in the southern extent of the Gulf of Mexico off the Yucatan peninsula. Initial computer models forecasted it would accelerate toward Florida. However, it slowly drifted northward. Therefore, NHC forecasters became skeptical as each subsequent model run projected a fast northward motion that did not materialize. Opal was also a Category 1 hurricane and was not expected to intensify past Category 2. Based on Opal's slow motion and the fact that it was a minimal hurricane, at 5 P.M. on October 3, 1995, NHC decided a hurricane warning was not necessary until the following morning for the Florida panhandle.

Late that night, Opal intensified explosively to a Category 4 hurricane with winds of 150 mph, just shy of Category 5 status. (Later analysis attributed this rapid development to movement over an isolated pool of very warm water. Interactions with environmental 40,000-foot winds may have also promoted the intensification.) Even worse, Opal finally began accelerating toward Florida. Because most people were sleeping, it was difficult to alert the public about this new, unanticipated danger. Many people near Pensacola Beach awoke the next morning with a near–Category 5 hurricane just offshore. The last-minute evacuation procedures clogged the roads and Interstate 10, creating traffic jams that moved less than 10 mph. It was NHC's worst nightmare coming true.

Fortunately, Opal weakened before landfall from a Category 4 to Category 3 hurricane. Although Opal was still dangerous and destructive, this reduction was a bit of a reprieve. Furthermore, although many residents were unable to use the roads and had to stay home during the storm, most residents on the immediate coast did manage to evacuate. Opal caused $3 billion in damage and killed nine people in Florida, Alabama, Georgia, and North Carolina, but none of these fatalities were from the storm surge. The people of Florida had dodged a bullet.

Opal demonstrated that large track forecast errors still occur, and when these errors are combined with unexpected intensification, many congested coastal regions are today toying with human catastrophe. Furthermore, seashore development has proceeded without seemingly any considerations for speedy evacuation. These issues will be discussed later, but first a background on storm motion and its prediction is necessary.

Factors Controlling Hurricane Motion

To understand hurricane motion, it is helpful to use the analogy of a wide river with a small eddy rotating in it. To a first approximation, the river transports the eddy downstream. However, the eddy will not necessarily move straight because the speed of the current varies horizontally; for instance, it may be faster in the center and slower toward the river banks. As a result, the eddy may wiggle off a little to the left or right as it moves downstream and, depending on the situation, may speed up or slow down. Furthermore, this eddy's rotation may alter the current in its vicinity, which in turn will alter the motion of the eddy.

Likewise, one may think of a hurricane as a vortex imbedded in a river of air. The orientation and strength of large-scale pressure patterns basically dictate the hurricane's motion, except that the steering depends both on the horizontal and vertical wind distribution. There is a tendency for stronger hurricanes to be steered by winds higher aloft, thus requiring the forecaster to identify the best "steering height" for a particular storm. To make a track forecast, one must first predict the large-scale flow within which the hurricane is imbedded, which can be difficult—how does one separate the hurricane winds from the environmental winds, and how does one know what height best represents the steering flow?

Forecast errors also occur when the steering current is weak and ill defined, thus causing the hurricane to stall or slowly drift. However, the largest errors are associated with rapid changes in the strength and orientation of the steering current. Under these situations, a hurricane may accelerate its motion and/or turn. For example, sometimes forecasters are faced with the dilemma of a hurricane moving toward the eastern coast of the United States, not knowing for sure whether it will hit land before being turned back to the right by westerly steering currents poleward of the hurricane.

The hurricane can internally change its course as well. Just as in the river example, the hurricane can interact with surrounding pressure fields, thus altering its own steering current. Research has also shown that the earth's rotation theoretically affects the motion by inducing a weak poleward and westward drift of 2–3 mph (known as the *beta effect*). On rare occasions when two hurricanes get within 850 miles of each other—especially in the Pacific and Southern Hemisphere oceans—they may begin to move toward each other and rotate about a common midpoint between them (known as the *Fujiwhara effect*). Even when a hurricane moves relatively straight, a detailed analysis shows small oscillations about the mean path, called *trochoidal motion.*

Computer Models

Until the late 1970s, NHC forecasters relied on meteorological intuition, statistical schemes, and the storm's past trend to make track predictions. They would also look at past tracks from other years to see which way similar storms moved. These techniques are still used today.

However, forecasters began to realize such difficult predictions required the use of *computer models.* Computer models take in current weather observations and approximate solutions to complicated equations for future atmospheric values, such as wind, temperature, and moisture. NHC uses a suite of models that differ in their mathematical assumptions and complexities in describing atmospheric processes. Some of these differences exist because certain atmospheric features have been removed to make the computer program run faster. Other differences exist because meteorologists are still uncertain about how to formulate certain weather features on a computer, such as cloud processes. The most complex models must be run on the fastest computers in the world, known as supercomputers. Model predictions can vary because of these mathematical differences. Model forecasts also contain small errors because the mathematical solutions contain a small amount of uncertainty that accumulates with time; therefore, all forecast errors grow the further into the future they are made. Models will also produce incorrect forecasts if they are initialized with bad data, which is a frequent problem over the data-sparse ocean (as they say, "garbage in, garbage out").

Nevertheless, computer models have revolutionized all aspects of meteorology, including hurricane track forecasts, and they are improving each year. In 1995, NHC started using a model developed at the Geophysical Fluid Dynamics Laboratory that is generally 20 percent better than its predecessors. Unfortunately, computer guidance for intensity forecasts is still unreliable. Forecasters use statistical schemes and intuition to make their intensity predictions.

Observation Platforms

Hurricane forecasters also monitor the latest observations from satellites, ships, radar, buoys, oil rigs, and other sources to assess any track or intensity changes. When a hurricane is far offshore in data-void regions, the intensity is estimated from satellite cloud patterns using the Dvorak technique. Although generally reliable, the Dvorak technique can produce wrong intensity values, as in the case of Typhoon Omar (1992). The Dvorak technique estimated Omar as a Category 1 storm, but it was discovered to be a Category 2 storm hours before hitting Guam, surprising the island's inhabitants and perhaps amplifying the damage due to lack of preparation.

Therefore, the most important data platform is *reconnaissance planes* that fly into the hurricane's eye and take critical meteorological measurements. The precise information required for obtaining hurricane structure, location, and intensity cannot be obtained in any other way, and this information is crucially important for forecasts and evacuation procedures. Forecasters credit a reconnaissance plane for discovering that Hurricane Camille (1969) had strengthened from sustained winds of 115 mph to 150 mph on August 16, 1969, at 5 P.M. CDT, 30 hours before landfall. Based on this new information, additional evacuations were prompted, which may have saved up to 10,000 lives. A reconnaissance plane also measured Hurricane Opal's (1992) unexpected rapid intensification. Reconnaissance flights are expensive, and they only occur in the Atlantic Ocean because only the United States government has the financial capability to fund them. At one time, Congress had considered halting the reconnaissance flights, but they reversed their position once the forecast and evacuation impacts were made clear by scientists and evacuation managers. Not only was funding preserved, a line item was added to the U.S. budget explicitly setting aside funds for Atlantic reconnaissance. At one time, Air Force reconnaissance flights also occurred in the western Pacific Ocean, but budget restrictions forced those flight missions to end in 1987. Occasional surprises like Typhoon Omar are the result.

Continuous reconnaissance flights begin once a tropical system moves close enough to land. The 53rd Weather Reconnaissance Squadron at Keesler Air Force Base near Biloxi, Mississippi—also known as the "Hurricane Hunters"—takes most of these measurements on WC-130 aircraft. The WC-130 can fly up to 15 hours at a cruising speed of 300 mph, and is configured with computerized weather instruments that measure hurricane motion, structure, and intensity. Two additional P-3 Orion planes with more sophisticated instruments (including a radar) are also available by the NOAA Aircraft Operations Center at MacDill Air Force Base in Tampa Bay, Florida. The P-3s are deployed on less routine flights to perform analysis of hurricane structure in conjunction with scientists at the Miami NOAA Hurricane Research Division (HRD). The P-3s also transmit data to NHC.

Obviously, reconnaissance flights carry an element of risk. In September 1955, a Navy plane and its crew of nine plus two Canadian newsmen were lost in the Caribbean Sea while flying in Hurricane Janet. Three Air Force aircraft have been lost flying in typhoons in the Pacific. Fortunately, no planes have been lost in the last 25 years, but scary moments still happen. In one extreme circumstance during Hurricane Hugo (1989), one of the P-3 planes encountered a severe mesoscale vortex, started sinking, and regained altitude moments before it would have crashed into the ocean. Most flights are less eventful, but do contain some bumps between thunderstorm updrafts and downdrafts, and occasional roller coaster-like drops (or ascents) of 3,000 feet in less than a minute due to strong updrafts and downdrafts. Because hurricane winds are strongest just above the surface (about 1,000 feet) and weaken with height, reconnaissance flights are performed between 5,000 and 10,000 feet. During the hurricane penetration, information about the horizontal wind and temperature structure is transmitted to NHC. Once the plane enters the eye, it deploys a tube of instruments (called a *dropsonde*) that parachutes downward from flight level to the sea, sending valuable intensity measurements back to NHC.

Summary of NHC Forecast Procedures

In summary, forecasters at the National Hurricane Center critically evaluate each computer prediction for reasonableness and consistency with other forecasts, observed data, statistical projections, the historical record, and extrapolated positions. They also monitor the latest observations for last-minute adjustments. Based on all of this information, every six hours the forecasters produce a new forecast of the storm's projected location and intensity for the next 72 hours. When hurricane conditions on the coast are possible within 36 hours, a *hurricane watch* is issued. When hurricane conditions are likely within 24 hours, a *hurricane warning* is issued. These watches and warnings refer to the arrival of hurricane force winds of 74 mph, not eye landfall, which generally occurs a few hours later. Because the margin of error for 24-hour track forecasts is currently about 100 miles, the warning is issued for a rather large coastal area (about 350 miles) and is quantified by the percent chance the hurricane's center will pass within 65 miles of a particular location.

Society's Preparation for Hurricanes— New Worries

Since Hurricane Camille killed 256 people in 1969, U.S. hurricane-related fatalities have dropped dramatically as storm surveillance, evacuation procedures, public awareness, and forecasts have improved. However, new concerns have emerged as coastal population growth and property development has exploded during this period. In fact, U.S. population increases are currently largest in coastal communities. The population has grown on average by 3–4 percent a year in hurricane-prone regions, and in some states like Florida the growth is greater. As a consequence, property damage costs due to hurricanes have skyrocketed in the 1990s (Table 1.8).

This trend is disturbing in other ways. First, during this period of larger property costs, the number of major hurricanes (Category 3 or higher) has actually decreased since 1970 (Landsea et al. 1996). This decrease corresponds with an active period of intense hurricanes during the 1940s to 1960s that suddenly changed to a quiet period of intense hurricanes during the 1970s

TABLE 1.8
Hurricane-Induced Property Damage Costs

Decade	Property Damage
1900–1909	$1.2 billion
1910–1919	$2.9 billion
1920–1929	$1.8 billion
1930–1939	$4.8 billion
1940–1949	$4.5 billion
1950–1959	$10.9 billion
1960–1969	$20.4 billion
1970–1979	$16.5 billion
1980–1989	$17.2 billion
1990–1995	$35 billion

Notes: Costs are adjusted to 1994 dollars.
Source: Hebert, P. J., J. D. Jarrell, and M. Mayfield, 1996. The Deadliest, Costliest, and Most Intense United States Hurricanes of this Century (and Other Frequently Requested Hurricane Facts). *NOAA Technical Memorandum NWS TPC-1.* Miami: National Oceanic and Atmospheric Administration, National Weather Service, National Hurricane Center, 30 pp.

to early 1990s. (Such shifts in weather activity are known as *multidecadal changes.*) This change is likely due to a cooling in the Atlantic Ocean water temperature during this period and may also be related to a drought that started in 1970 in Africa, which may have weakened tropical waves before propagating out into the Atlantic.

Second, the number of major hurricanes making landfall on the eastern coast of the United States has dramatically decreased since 1965. During the period 1944 to 1964, 17 major hurricanes hit the East Coast (Landsea 1999). In contrast, only 5 major hurricanes made landfall on the East Coast between 1965 and 1995 (with a period between 1965 and 1983 where *no* East Coast landfalls occurred!). During this period much population growth and development occurred on the East Coast, apparently ignoring or not knowing what happened between 1944 and 1964.

Some evidence suggests that major hurricane activity occurs in 40–60 year cycles, where 20–30 years of active Atlantic hurricane seasons will be followed by 20–30 years of relatively quiet hurricane seasons (Gray et al. 1996). In fact, there are signs that the downward trend in hurricane activity may be ending. The year 1995 was extraordinarily active, with 19 named storms (including both tropical storms and hurricanes). Of these 19 storms, 11 were hurricanes, and 5 of the 11 were major hurricanes (Category 3 or stronger). The years 1996 and 1998 also had above normal storm activity. In fact, the four-year period of 1995–1998 had a total of 33 hurricanes—an all-time record in the Atlantic Ocean. Global weather patterns are also emerging that are favorable for more active hurricane seasons in the future, such as a return of warm Atlantic water temperatures (Gray et al. 1998). When the more active hurricane phase returns (if it hasn't already), property damage costs could easily exceed $50 billion in one year, especially should a major city be hit. Regardless of trends in Atlantic hurricane season activity, each year the United States has at least a one in six chance of experiencing hurricane-related damage of at least $10 billion (in 1995 dollars) (Pielke and Landsea 1998).

Even more disturbing is the sharp population increase. Most new coastal residents have never experienced a hurricane, increasing the chances some may ignore evacuation orders or not take proper precautions. Also, some coastal regions are so congested now it is more difficult to perform a timely evacuation. Track forecasts have improved by about 1 percent per year, but the 3–4 percent population growth could overwhelm these better predictions, resulting in higher casualties again. Even worse, should a hurricane unexpectedly change course or accelerate its motion, there may not be enough lead time for orderly evacuation of intensively developed coastal regions, trapping many residents at landfall.

Even before a hurricane makes landfall, millions of dollars are lost in evacuation costs. At least $300,000 of business losses and hurricane preparation costs is incurred per day for every mile of coastline evacuated. Some experts claim the cost is even larger, perhaps $1 million per mile.

Therefore, a top priority has become accelerating the improvements in forecasting accuracy. Research has shown that more detailed reconnaissance observations of hurricanes may achieve this goal. As a result, HRD has improved the dropsonde using Global Positioning System technology so that hurricane observations may be obtained with unprecedented accuracy. In addition, the new dropsondes have the ability to take observations below 1,500 feet—something the older dropsondes could not do. A new, faster plane has also been added to the fleet—the Gulfstream IV. Because the Gulfstream is a jet, it can reach altitudes of 45,000 feet, unlike older propeller-driven planes belonging to the Hurricane Hunters and to the Aircraft Operation Center, which can only reach 25,000 feet at best.

In 1998, the Gulfstream IV deployed dozens of dropsondes per flight at 40,000 feet around several hurricanes, and similar data sampling strategies will be used in the future. Such efforts will relay a better picture of the atmospheric conditions that surround and steer a hurricane. Other Air Force and HRD planes deployed drifting buoys in the path of hurricanes to better understand storm conditions near the ocean surface. Research has shown that this additional data can reduce computer model forecast track errors up to 30 percent.

If this estimated track improvement is correct, it may allow NHC to shrink the warning zone by 50 to 80 miles, which would save millions of dollars in evacuation costs, especially if a major city is excluded from the warning area. It would also reduce traffic congestion, speeding up evacuation from the most threatened region. Furthermore, if homeowners take advantage of more accurate forecasts by seriously protecting their homes, it is estimated another 10–15 percent in costs would be saved—and any lives saved by the improved forecasts would be priceless.

Other exciting technological developments are about to occur. The Hurricane Hunters will be replacing their WC-130H aircraft with WC-130J planes in 1999; the latter are faster, more fuel efficient (resulting in longer flights), and most importantly can travel at altitudes of 37,000 feet or more. These improvements will further augment operational measurements critical to hurricane forecasting. In addition, starting in 1998 NASA scientists, NOAA scientists, and universities are collaborating to collect data at all levels of hurricanes using multiple aircraft, remote sensing technologies, and portable surface measuring platforms (see Chapter 2 for details about the Third Convection and Moisture Experiment, also called CAMEX3). The data sets produced by this and future missions will be unprecedented in their comprehensiveness, providing researchers with new information toward understanding hurricanes.

By the twenty-first century, another new type of plane may be taking oceanic observations. Scientists in Australia and Canada are experimenting with small, pilotless aircraft typically called "drones," or sometimes "aerosondes." These drones theoretically can be programmed to fly a fixed path for over 24 hours, continuously taking weather observations that would improve weather forecasts. The first successful cross-Atlantic flight by an unmanned aerosonde was accomplished in August 22, 1998. Their unique aerodynamic structure and extremely light weight also allow them to theoretically withstand hurricane winds and strong updrafts in thunderstorms, thereby providing timely and continuous crucial weather data at less cost than airplane reconnaissance.

Forecasting Annual Hurricane Activity

In the early 1980s, Dr. William Gray at Colorado State University asked the question, "Why do some Atlantic hurricane seasons have many storms and why are other seasons inactive?" He further wondered if one could predict, months in advance, the number of tropical storms and hurricanes for the upcoming Atlantic tropical season. Based on his research, in 1984 Gray began to publicly predict how active the Atlantic hurricane season would be before it starts. His predictions, at times, have been remarkably accurate and have scientifically proven to be skillful compared to guessing. His forecasts, which are well publicized in the media, are issued every December, April, June, and August. The forecast is also available on the web at http://tropical.atmos.colostate.edu.

Gray and his students have discovered several surprising global signals that affect Atlantic hurricane activity (Gray et al. 1998). The reasons for some of these associations remains unclear and are still being researched. The signals include the following:

1. *The El Niño-Southern Oscillation (ENSO)*, also called simply *El Niño*. El Niño occurs every 3–7 years when warm water in the equatorial western Pacific Ocean shifts east. What has the Pacific Ocean got to do with Atlantic hurricanes, you ask? This change in the Pacific Ocean alters global wind patterns and modifies many weather elements, including reducing hurricane activity. In El Niño years, wind shear is enhanced in the Atlantic Ocean, which is destructive to tropical storm and hurricane formation. An El Niño officially ends when the warm eastern Pacific waters cool to average temperatures; however, usually the cooling continues, which is called a *La Niña*. La Niña years are associated with weaker than average Atlantic wind shear and enhanced Atlantic hurricane activity. The occurrence of an El Niño or La Niña is one of the most important modulators of Atlantic hurricane activity.

2. *African rainfall.* There is a strong correlation between rainy years in the western Sahel of Africa and major (Category 3, 4, or 5) hurricane activity in the Atlantic. Likewise, during drought years in Africa major hurricane activity is reduced. Most major hurricanes come from tropical waves that originate over Africa. It has been hypothesized that, during rainy years, tropical waves are stronger and more conducive to genesis as they move westward off the African coast. However, Goldenberg and Shapiro (1996) show that years with strong wind shear are associated with droughts in Africa and years with weak wind shear, with rain. African rainfall is also likely related to changes in water temperature in the Atlantic Ocean, where years with warm water are correlated with wet weather in Africa and more strong hurricanes, and years with cold water are correlated with dry weather in Africa and fewer strong hurricanes (C. Landsea, personal communication 1998).

3. *Pressure and temperature difference between the western African coast and the Sahel region during the previous February-May period.* When these differences are conducive to onshore flow that transports moisture over Africa, hurricane activity is often enhanced; when it is conducive to offshore flow, hurricane activity is often suppressed.

4. *Caribbean sea surface pressure.* When pressure is lower than normal in the Caribbean Sea, Atlantic hurricane activity is enhanced; when it is higher than normal, hurricane activity is suppressed. High pressure indicates sinking air, which suppresses thunderstorm formation in easterly waves and other tropical disturbances. Knaff (1997) showed that higher pressure is also correlated with strong wind shear.

5. *Quasi-Biennial Oscillation (QBO).* The QBO is an oscillation of equatorial winds between 13 and 15 miles aloft. These winds change direction between westerly and easterly every 12–16 months. Westerly winds are associated with more hurricanes than easterly, especially when the wind is westerly at both 13 and 15 miles aloft. It has been postulated that a westerly QBO reduces wind shear, but the reason for this hurricane association remains unclear.

6. *Caribbean wind shear.* During seasons when shear is greater than normal in the Caribbean Sea, hurricane activity is suppressed, and vice versa for seasons with weak wind shear.

7. *Atlantic Ocean water temperature.* Obviously, warmer than normal sea surface temperature is favorable for active hurricane years, and vice versa for colder than normal water.

8. *Strength of Azores High Pressure System.* The Azores High is a permanent feature located between 20 and 30 degrees W in the Atlantic Ocean. When pressure is higher than normal in the Atlantic, tropical northeast winds tend to be stronger than normal. This wind, in turn, upwells cold water to the surface off the northwest African coast, thereby reducing the chance of tropical genesis as tropical waves propagate off Africa. There also is some long-term feedback, such that a strong Azores High in the fall results in high pressure in the Caribbean Sea the

following year (Gray et al. 1998; Knaff 1998). Therefore, when the Azores High is stronger than normal in the central Atlantic, hurricane activity is reduced; likewise, when the Azores High is weaker than normal, hurricane activity is enhanced.

In general, when more of these predictors are favorable for hurricane activity than unfavorable, Gray predicts an above average hurricane season, and when most are unfavorable, a quiet hurricane season is predicted. When the same number of predictors have positive and negative influences, an "average hurricane season" is predicted (about nine named storms, with six of these named storms becoming hurricanes and two becoming intense hurricanes; see Table 1.1). However, caution must be advised because some predictors are more important than others. For instance, despite the fact that most factors were favorable for an active hurricane season in 1997, a record El Niño that year dominated the weather, resulting in a quiet hurricane season. Also, one should be careful attributing Atlantic hurricane activity to one feature in any year, because many of these features are interrelated; for example, an El Niño tends to be associated with strong Caribbean wind shear.

Gray quantitatively predicts several parameters using statistical techniques and intuition, such as the number of named storms (which includes tropical storms and hurricanes), the number of hurricanes, the number of days with tropical storms or hurricanes, and so on. He also forecasts the number of major hurricanes, which includes Category 3, 4, or 5 hurricanes, since these are the storms that cause about 83 percent of total hurricane damage (Pielke and Landsea 1998). Table 1.9 summarizes Gray's forecast of named storms versus the observed number.

Note from Table 1.9 that during 11 of 15 years, Dr. Gray correctly predicted whether the observed number of storms would be above or below the average of nine named storms. However, the years that were incorrect (1989, 1993, 1997, and 1998) are instructive because they led to new insight on how to improve the forecasts. For example, after analyzing the 1989 underforecast of storm activity, the importance of African rain was discovered because 1989 was the only nondrought year that decade. Starting in 1990, African rainfall was included in Gray's seasonal forecasts. Also, because the predictions are based on

TABLE 1.9
Verification of Dr. Bill Gray's Seasonal Hurricane Forecast

Year	Number of Named Storms Forecasted	Number of Named Storms Observed
1984	10	12
1985	10	11
1986	7	6
1987	7	7
1988	11	12
1989	9	11
1990	11	14
1991	7	8
1992	8	6
1993	10	8
1994	7	7
1995	16	19
1996	11	13
1997	11	7
1998	10	14

Note: Number of named storms in the Atlantic (including both tropical storms and hurricanes) predicted by Dr. Bill Gray each year in August, versus the number actually observed.
Source: Gray, W. M., P. W. Mielke, Jr., and K. J. Berry. 1998. Extended Range Forecast of Atlantic Seasonal Hurricane Activity and U.S. Landfall Strike Probability for 1998. Department of Atmospheric Science Paper, Colorado State University. Also available at http://tropical.atmos.colostate.edu.

statistical probabilities, the forecast will fail in some years (just as local National Weather Service statistical forecasts of rain and temperature will occasionally fail). Gray has to also make a correct prediction of such parameters as African rain and El Niño for an accurate seasonal hurricane forecast. If he forecasts a basic predictor incorrectly, the seasonal forecast will probably be wrong. For example, Gray's 1997 hurricane prediction was incorrect because no one anticipated the unusually strong El Niño that occurred that year.

Starting in 1998, Gray will begin issuing forecasts for hurricane landfall probability (Gray et al. 1998). It will be interesting to observe how this new extended-range forecast methodology will evolve.

All hurricane seasons should be taken seriously by coastal residents, even if the seasonal forecast is "below normal." Many devastating hurricanes have occurred in otherwise inactive seasons, such as Hurricane Andrew (1992), Hurricane Alicia (1983), and Hurricane Allen (1980).

Is Global Warming Increasing the Number of Hurricanes?

Some environmental groups claim hurricanes are becoming more numerous and stronger (Leggett 1994) due to a reputed phenomenon known as "global warming." Media outlets also have made such assertions (Newsweek 1996). Before considering this claim, one must realize that the topic of global warming is itself very political and controversial. Although a majority of scientists believe global warming is occurring, many scientists still discredit these allegations with solid arguments. This book will not discuss the global warming controversy in any detail because many other books are devoted to the subject, including an ABC-CLIO book (Newton 1993).

However, two issues concerning any links to possible global warming and hurricanes will be discussed. The first question is: Have hurricanes increased in numbers and/or intensity in the past two decades because of possible global warming? The second question is: Will hurricane activity increase in the future because of possible global warming?

In short, the answer to the first question is: no. The answer to the second question is: maybe, although reductions in hurricane activity are also possible. Before addressing these questions in detail, a review of the greenhouse effect and global warming is required.

What Is the "Greenhouse Effect"?

A balance between incoming solar radiation from the sun and outgoing radiation from the earth primarily regulates the atmosphere's temperature. This is a complicated process requiring some explanation. When an object becomes warmer than $-273.15°C$ (the temperature where all molecular motion ceases, also called *absolute zero* on the Kelvin temperature scale, and equivalent to $-459.67°F$), it begins to emit electromagnetic radiation. Therefore, as the sun warms the earth environment, the earth begins to emit its own radiation. However, radiation is dependent on temperature. The hotter an object is, the shorter the wavelength of peak emission. The much hotter sun emits most of its radiation as visible light. The cooler earth emits its energy at a much longer wavelength that human eyes cannot see called the infrared spectrum.

It turns out that molecules in the atmosphere behave differently with regard to the passage of infrared radiation and visible radiation through them. It is a complex process, but the end result is that some air molecules allow visible radiation to pass through them but obstruct infrared radiation. With this background, the heat balance of the earth and its atmosphere can now be discussed.

As solar radiation flows from the sun toward the earth, it passes freely through the atmosphere and warms the earth. However, as the earth tries to emit this heat back to space in the infrared spectrum, some of it becomes "trapped" by the atmosphere, thus warming the air. A portion of this energy is radiated back to the earth, which warms the surface. The earth, in turn, reradiates this infrared energy upwards, where it is again absorbed and warms the lower atmosphere some more. In this way, the atmosphere acts as an insulating layer, keeping part of the infrared radiation from escaping rapidly into space. Consequently, the earth's surface and lower atmosphere are much warmer than they would be without this selective absorption of infrared radiation. This process is often called the *greenhouse effect* because it is analogous to a glass building that allows visible light inside but prevents some infrared radiation from leaving, thus keeping the plants inside warm (even in winter). It is important to realize that the atmosphere's greenhouse effect is a natural process, and without it the earth would be a much colder, unlivable planet with an average surface temperature of 0°F.

There is a distinction between the atmosphere and a greenhouse—in a greenhouse the glass allows visible light to pass through but restricts the passage of infrared, whereas in the atmosphere molecules are differentiating between visible and infrared radiation. Also, a greenhouse warms quickly because glass is a physical barrier to air movement, whereas in the atmosphere mixing occurs freely between warm air at the ground and cooler air aloft, thus slowing down warming. Despite these fundamental differences, the phrase "greenhouse effect" is used throughout the media and meteorology, so the nomenclature will be used in this book as well.

Certain molecules are more effective at trapping infrared radiation than others. The most important greenhouse gas is water vapor because it strongly absorbs a portion of outgoing infrared radiation and is a plentiful gas. The other greenhouse gas is carbon dioxide (CO_2), which is equally as absorptive but far less plentiful than water vapor and therefore plays a much smaller role in the greenhouse effect. CO_2 is a natural component of the atmosphere, produced mainly by the decay of vegetation.

What Is "Global Warming"?

CO_2 is also produced by the burning of fossil fuels, such as coal, oil, gasoline, and natural gas. Observations show that CO_2 has increased by more than 10 percent since 1958, coinciding with increases in fossil fuel emissions from automobiles, factories, and other power sources. *Global warming theory* states that the greenhouse effect will be enhanced as CO_2 increases because more outgoing infrared radiation will be absorbed. There is also the possibility that increasing the temperature will also increase water vapor concentrations (the major greenhouse gas) due to increased evaporation rates, which would further enhance the warming prospects. Some scientists contend global warming will increase the earth's average surface temperature by 2.5° to 8°F. However, other scientists contend feedback processes involving clouds and the ocean may cancel that potential warming. Some scientists, based on decades of observations, also say global warming is already occurring. Other scientists using different observation techniques, however, show no true warming has happened yet; others attribute any perceived warming to natural climate variability. The honest truth is no reputable scientist is 100 percent confident that global warming has occurred or that it will ever occur. But let's play devil's advocate for the moment and discuss how these scenarios could change hurricane activity.

Have Hurricanes Increased in Number or Strength Due to Potential Global Warming?

The global average of 86 tropical storms has probably not changed in the last few decades (Landsea 1999). However, some ocean basins do experience 10- to 20-year cycles in which the total number of storms changes. For example, since 1980 the total number of tropical storms has increased in the Northwest Pacific, but this increase was preceded by a nearly identical decrease from about 1960 to 1980. A downward trend in tropical storm number since the mid-1980s for the Australian region has been observed, but this is probably an artificial decrease due to a change in tropical cyclone wind designation by that country in the mid-1980s.

As discussed earlier, the number of Atlantic tropical storms has substantial year-to-year variability. However, no significant trend in tropical storm number has been observed since 1944 (Landsea et al. 1996). In contrast, the number of major hurricanes (Category 3 or higher) has shown a significant *downward* trend during this period, corresponding to an active period of intense hurricanes during the 1940s to 1960s, which suddenly changed to a quiet period of intense hurricanes during the 1970s to 1990s. (Such shifts in weather activity are known as multidecadal changes). This change is likely due to a cooling in the Atlantic Ocean water temperature during this period and may also be related to a drought that started in 1970 in Africa, which causes many tropical waves to form before propagating out into the Atlantic.

The early 1990s were a particularly inactive period in the Atlantic. No hurricanes were observed over the Caribbean Sea during the years 1990–1994, which was the longest period lacking hurricanes in the area since 1899. The period 1991–1994 is the quietest four-year period on record since 1944 in terms of frequency of total storms (7.5 per year), hurricanes (3.8 per year), and major hurricanes (1.0 per year). However, one major hurricane making landfall during this period was Andrew (1992), which caused a record amount of damage in Miami.

Because Andrew, a Category 4 hurricane, hit a metropolitan area, it attracted much attention, including claims that global warming must have caused such a powerful storm because of the unprecedented damage (Leggett 1994; Newsweek 1996; U.S. Senate Bipartisan Task Force on Funding Disaster Relief 1995; Dlugolecki et al. 1996). However, it is normal to have two major hurricanes each year (Table 1.1). In fact, it is unusual to have only one major hurricane per year during a four-year period! Andrew just happened to hit a highly developed coastal region.

Therefore, claims of global warming increasing hurricane activity or intensity in the last few decades are incorrect. In fact, during the last 50 years the overall trend has been a *decrease* in major hurricanes during this period of increased fossil fuel emissions. If global warming is already occurring (and no one knows this for sure yet), thus far it definitely has not increased hurricane activity.

The moral is that a reader should view most sensational science reports with some skepticism. Stories about global warming (and most sensational science stories in general) need not necessarily be dismissed as incorrect, but they should be carefully scrutinized. Questions one should ask when preparing a science report, or trying to learn about a particular science topic, are:

1. *Who is the source of the report?* Is it in a scientific magazine written by an author with a science degree from college, or is it in a newspaper or television report by an author with no background in the sciences? Just because an article is in popular media outlets does not mean it is shoddy, but its accuracy does depend on the quality of the journalist, his or her interview procedures, and his or her scientific aptitude. Some writers work hard to produce an accurate, informative article. Many scientists, however, can describe instances in which journalists inaccurately reported their information or they had quotes taken out of context.

2. *Was the report peer-reviewed?* The most reputable reports are in journals that undergo a peer-review process. The peer-review process involves rigorous commentary by several experts in the field who either accept or reject the article based on its scientific merit. Usually several rewrites are needed to make an article acceptable. Examples of general science journals available at bookstores and newsstands that at least undergo some minimal peer review are *Scientific American, Nature,* and *Science.* Examples of peer-reviewed journals in meteorology are *Monthly Weather Review, Weather and Forecasting, Journal of Climate,* and the *Bulletin of the American Meteorological Society.* The latter are all published by the American Meteorological Society (see Chapter 5).

3. *Is it a well-balanced report presenting all the issues, or is it strongly biased toward one point of view?* Often journalists, or magazines funded by a particular special interest group or political party, will strongly skew a report toward their own point of view. These reports may have legitimate information, but one should investigate other sources before jumping to conclusions. To assess whether a report is biased, simply look at other reports by the same reporter or magazine to see if similar topics are reported in the same context. Even peer-reviewed articles can contain biases.

In summary, readers should be skeptical and inquisitive about most controversial or sensational scientific information they encounter, and they should always obtain their information from more than one source.

If Global Warming Occurs in the Future, Will Hurricanes Increase in Numbers and Intensity?

Research has shown that global warming could increase tropical sea surface temperatures and tropical rainfall. Based on this research, some have suggested hurricanes may increase in frequency, area of occurrence, and intensity (Ryan et al. 1992; Emanuel 1987). However, any changes in hurricane activity will also be associated with large-scale changes in the tropical atmosphere (Landsea 1999; Henderson-Sellers et al. 1998). For example, the instability threshold necessary for thunderstorm maintenance could change, because temperature may increase throughout the atmosphere, not just at the surface.

In addition, any changes in global wind patterns would profoundly affect hurricane activity. For instance, if global warming increases wind shear, one would see a significant decrease in hurricane activity. Likewise, a reduction in wind shear would dramatically increase hurricane activity. If monsoon activity increased, then tropical cyclogenesis would increase, and vice versa for weakened monsoon troughs. Another wildcard is how global warming would change El Niño activity—if more El Niños occurred, hurricane activity would be reduced, and vice versa for fewer El Niños. In summary, the combined changes in water temperature, wind shear, monsoon activity, atmospheric instability, and El Niño would dictate how hurricane activity changes. Therefore, it is difficult to assess how potential global warming could alter hurricane activity. In any case, there is considerable motivation for society to prepare better for hurricanes independent of global warming concerns!

Attempts at Hurricane Modification

Because of their destructive and life-threatening nature, experimental attempts have been made to weaken hurricanes. The main hypothesis involves converting liquid cloud water to ice just outside the eyewall. Water gives off enormous quantities of stored heat (also called *latent heat*) when it changes phase from liquid to ice. Many clouds exist in a *supercooled* state, which means the liquid droplets' temperature is below freezing (32°F) but the liquid lacks a "triggering" mechanism to turn to ice. For this conversion to occur, supercooled liquid water needs to attach to a floating aerosol with a molecular structure similar to ice, known as *ice nuclei*. However, ice nuclei are often sparse in the atmosphere, and many supercooled droplets are never converted to ice. If somehow one could introduce artificial ice nuclei (such as silver iodide) into a supercooled cloud, the water would be converted to ice, thus releasing lots of heat into the air and causing the cloud to grow. This premise, known as *cloud seeding,* has been attempted to increase rainfall in drought-stricken regions (with unproven results) and snowpack in the mountains (with successful results).

To theoretically weaken a hurricane, the seeding process is more complicated. At first scientists thought they could seed the eyewall and perturb the hurricane's wind outward, thus weakening the hurricane (Anthes 1982; Willoughby et al. 1985). By the mid-1960s scientists realized this theory was flawed. A revised theory evolved in which one seeds the clouds just outside the eyewall to stimulate cloud growth away from the eyewall. The new outer eyewall would grow, depriving inflow into the older, inner eyewall. The result is a weakening inner eyewall, resulting in less subsidence in the eye and a rise in central pressure. If the pressure increases, inflowing air is unable to penetrate to as small a radius, and most of the new ascent occurs at the new outer eyewall. Eventually the new eyewall would replace the old eyewall, but at a larger distance from the center. Just as ice skaters slow their rotation when their arms are spread out, a larger eyewall radius would cause a reduction in wind speed.

Cloud seeding was first tested when several U.S. government agencies collaborated in a pioneering weather modification effort known as *Project Cirrus* (Willoughby et al. 1985). Among other notable firsts was the first cloud seeding of a hurricane. On October 13, 1947, a plane dropped silver iodide into a hurricane moving to the northeast. Observers on the plane noted changes in the visual appearance of the cloud but could not demonstrate any changes in structure or intensity. However, shortly afterward the hurricane reversed course to the west, making landfall on the coasts of Georgia and South Carolina. It is extremely unlikely the seeding altered the course, because hurricanes are mostly guided by constantly shifting large-scale atmospheric currents. However, the political and legal implications taught scientists to be more careful about where they conducted their hurricane seeding experiments. Future attempts at hurricane modification occurred only in hurricanes that were (1) far from all land masses, and (2) unlikely to make landfall within 24 hours.

The next experiment occurred on September 16, 1961, when a naval aircraft dropped eight canisters of silver iodide into Hurricane Esther. Esther, which had been intensifying, stopped strengthening and the eyewall's distance from the center increased. The next day, a second seeding attempt was made, but the canisters missed the eyewall and the hurricane's intensity did not change. At the time, the experiment was considered successful, although we now know that the eyewall-seeding hypothesis was flawed and that Esther's changes must have occurred naturally. The encouraging results from Esther led to the formal establishment of a hurricane modification program known as *Project STORMFURY* in 1962.

STORMFURY was a collaboration between the National Oceanic and Atmospheric Administration and the U.S. Navy and was directed by the National Hurricane Research Project (now called the Hurricane Research Division). STORMFURY started out well when Hurricane Beulah's eyewall was seeded on August 24, 1963. The eyewall disintegrated, followed by formation of a new eyewall at a larger distance from the eye. The maximum winds decreased by 20 percent and moved further away from the center. STORMFURY seemed to have a promising beginning.

Four more years were to pass before the next modification experiment. The years 1964–1968 were generally inactive hurricane seasons, and the hurricanes that did occur were either too close to land or out of flight range. Scientists also realized that the eyewall-seeding hypothesis was incorrect, resulting in the revised hypothesis of seeding outside the eyewall. The modified hypothesis was tested on Hurricane Debbie on August 18 and August 20, 1969, when more than a thousand seedings of silver iodide were made each day. The eyewall shifted outward on each of these days, and the winds decreased by 31 percent and 15 percent, respectively.

Ironically, this seeding was essentially the end of Project STORMFURY (Willoughby et al. 1985). The 1970 hurricane season yielded no suitable candidates for seeding. In 1971, the only eligible storm was Hurricane Ginger, a late-season, diffuse system. It was seeded twice but was a poor candidate because it lacked a small, well-defined eye; the seeding had no effect on Ginger. In 1972, all the storms were too weak, too close to land, or out of flight range. In fact, in general the 1970s were a period of below average hurricane activity, especially for intense hurricanes. Other difficulties also plagued STORMFURY. The Navy ended its support to pursue goals more closely related to national defense. Several of the aircraft had become too old for reliable use. Permission for seeding from Caribbean countries and Mexico became more difficult to obtain, and the State Department increased the time restriction for seeding before landfall (Posey 1994). Eventually, only a narrow zone north of Puerto Rico in which hurricanes were at least 36 hours from landfall was allowed for seeding. Unfortunately, no storms ever passed through the small permissible trapezoid of ocean north of Puerto Rico between 1973 and 1979. STORMFURY scientists attempted to move the project into the Pacific, where storms are more numerous, but Japan and Australia blocked that move. Finally, weather modification experiments in general were falling into public disfavor (Pielke and Pielke 1997). In 1983, Project STORMFURY was terminated.

Also, from the very beginning of STORMFURY and throughout the project's lifetime, several scientists expressed concerns about the seeding hypothesis, as well as the interpretation of STORMFURY's experiments. These doubts became substantiated by additional reconnaissance observations in the 1970s and 1980s and by increased knowledge of hurricane structure and evolution during this period. These findings, which refute the basic premise of project STORMFURY, are summarized as follows (Willoughby et al. 1985):

1. Observational evidence shows that hurricanes actually contain too little supercooled water and too much natural ice for seeding to be effective.
2. Hurricanes go through a natural (but temporary) weakening process in which a new eyewall forms outside the original eyewall. The outer eyewall "chokes off" inflow to the inner eyewall, causing it to dissipate. The outer eyewall then propagates inward, replacing the original eyewall. This process, known as the *concentric eyewall*

cycle, lasts 12 to 36 hours and is associated with temporary weakening (Willoughby et al. 1982). Therefore, much of the observed weakening in Esther, Beulah, and Debbie may be the result of natural internal evolution rather than a consequence of seeding. Hurricane intensity is also controlled by external influences, such as water temperature and wind shear. For example, some evidence exists that Debbie moved into a strong wind shear environment conducive to weakening (P. Black, personal communication 1994). Therefore, the expected results of seeding are indistinguishable from naturally occurring intensity changes.

Project STORMFURY should not be viewed as a failure, however. Even though only a few seeding experiments were performed, many reconnaissance flights were conducted into hurricanes to understand their formation, structure, and evolution. The legacy of Project STORMFURY is a wealth of observations that have augmented our understanding of hurricanes and have improved hurricane forecasts. For more information about Project STORMFURY, the reader is referred to papers by Posey (1994) and Willoughby et al. (1985).

The notion of hurricane modification still remains (Posey 1994). Occasionally, someone will ask, "Why don't we just nuke a hurricane and destroy it?" Radioactivity aside, such a question demonstrates an extreme underestimate of hurricane power. For example, Hurricane Andrew (1992) generated the equivalent energy of a 10-megaton bomb continuously during its existence—not in a split second as in a bomb explosion! It is doubtful a hurricane would even be affected by a nuclear blast, other than spewing radioactive fallout throughout the region. Other ideas involve removing a hurricane's energy source. Some have proposed covering the ocean ahead of a hurricane with an impermeable chemical film that impedes evaporation from the sea, thus weakening the hurricane as it moves into that region (Posey 1994). This proposal is unrealistic because it is unlikely any surface film could withstand the 30- to 50-foot ocean waves in a hurricane and because covering hundreds of square miles of ocean with a substance is a formidable task. Others have promoted stimulating cloud growth in the outer core of a hurricane by increasing the surface temperature with carbon black, which possibly would restrict inflow to the hurricane's inner core (Gray and Frank 1993);

however, such a scheme would suffer the same obstacles as Project STORMFURY. Besides, both schemes pose environmental problems. The wisest course of action is to reduce societal exposure to the effects of hurricanes, not through modification attempts. Pielke and Pielke (1997) offer several recommendations to improve hurricane preparedness policies.

Summary and Future Outlook

This chapter has provided a general overview of hurricane formation, structure, and destructive capability. The procedures, difficulties, and inaccuracies in forecasting these devastating storms have been presented. Processes controlling active and inactive Atlantic hurricane seasons have been described. The global warming controversy has been discussed, showing that hurricane activity has not increased in the last 50 years due to global warming. The history of a hurricane modification program, known as Project STORMFURY, was described from its inception to its termination when its results proved inconclusive and its theoretical basis proved flawed.

It is important that coastal residents never underestimate hurricanes. Between 1970 and 1994, fewer landfalling major hurricanes occurred than in the 1950s and 1960s. During this inactive period, U.S. coastal population and property development has exploded. When landfall did occur between 1970 and 1994, the result has been sky-rocketing damage costs, increasingly difficult evacuation procedures, and sometimes complacency from coastal residents who have never experienced a hurricane. For example, many Florida Key residents did not evacuate when Hurricane Georges (1998) threatened the area. Had Georges been stronger than a Category 2, many Florida Key residents would have been killed.

The future outlook is even more ominous. Some evidence suggests that the Atlantic Ocean experiences decadal fluctuations in water temperature, with some decades experiencing warmer water than normal, and other decades abnormally cold. These water temperature changes are linked to slow salinity changes in the North Atlantic known as the "Atlantic Conveyor Belt." These decadal cycles tend to correspond with Atlantic hurricane activity. During the period of decreased Atlantic hurricane activity between 1970 and 1994, Atlantic waters were colder than normal. However, in 1995 Atlantic waters switched from a cold phase to a warm phase. The next four years—1995–1998—were the most active consecutive four-year period on record, with 53 named storms, 33 hurricanes, and 15 major hurricanes (Category 3, 4, or 5).

In other words, due to this climate shift, Atlantic coastal residents may experience an increasing number of landfalling major hurricanes in the next 20 years. Major hurricanes cause 80 to 90 percent of all hurricane-linked destruction, and it is quite possible the United States will experience unprecedented property damage during this period. Just one major hurricane, should it hit a metropolitan region, could easily double or triple the record $26.5 billion damage inflicted by Hurricane Andrew in 1992.

The rest of this book provides background hurricane information for those interested in a better understanding of these storms. Enclosed in this text are a chronology of hurricane events and discoveries, biographies of influential hurricane scientists and forecasters, relevant statistics, excerpts on controversial topics, testimonials, hurricane preparedness tips, disaster recovery tips, and a list of books and electronic resources. A description of relevant organizations that are involved in hurricane forecasting, research, and mitigation is also provided in one chapter. A glossary is enclosed that contains definitions, additional tables, a hurricane tracking chart, and a description of NHC forecast bulletins. The following chapters should provide insightful information for any reader, whether he or she is performing historical research, looking for more information on a specific subject or storm, investigating hurricane preparedness procedures, or just trying to learn more about hurricanes.

Notes

1.   It is a popular belief that modern Australians call hurricanes *willy-willies*. The word was coined by a colonial settler sometime in the early 1900s and was only used in northwest Australia in reference to hurricanes. However, this practice ended over 50 years ago. Further confusing matters is that in the southeast of the country in the latter half of the twentieth century the same term (willy-willy) was used to describe dust devils. A dust devil is a small, short-lived whirl of dust, sand, or debris created by strong solar heating that promotes rising columns of air near the ground—definitely a different phenomenon than a hurricane.

2. For more information about monsoon depressions, see Ahrens (1994). For more information about subtropical cyclones, see the glossary section of the National Hurricane Center web site at www.nhc.noaa.gov.

3. This scale was devised in 1971 by Herbert Saffir, an engineer in Miami, for the World Meteorological Organization and was given to NHC. Robert Simpson, the director of NHC, then added the storm surge portion.

References

Many of the articles listed below are from meteorological journals. A description of the journal abbreviation is provided in the appendix.

Ahrens, C. D., 1994. *Meteorology Today.* St. Paul, MN: West Publishing Co., 591 pp.

Anthes, R., 1982. *Tropical Cyclones: Their Evolution, Structure, and Effects,* American Meteorological Monographs, Volume 19. Boston: American Meteorological Society, 208 pp.

Corps of Engineers, 1970. *Hurricane Camille—14–22 August 1969.* U.S. Mobile, AL: Army Engineer District, 80 pp.

Danielson, E. W., J. Levin, and E. Abrams, 1998. *Meteorology.* Boston: WCB/McGraw-Hill Co., 462 pp.

DeMaria, M., and J. Kaplan, 1994. Sea Surface Temperature and the Maximum Intensity of Atlantic Tropical Cyclones. *J. Climate,* 7:1324–1334.

Dlugolecki, A. F., K. M. Clark, F. Knecht, D. MacCauley, and W. Yambi. 1996. Financial Services. In *Climate Change 1995: Impacts, Adaptations, and Mitigation of Climate Change: Scientific-Technical Analyses.* Contribution of Working Group II to the Second Assessment Report of the Intergovernmental Panel on Climate Change. Cambridge: Cambridge University Press, Chapter 17.

Dvorak, V. F., 1975. Tropical Cyclone Intensity Analysis and Forecasting from Satellite Imagery. *Mon. Wea. Rev.,* 103:420–430.

Emanuel, K. A., 1987. The Dependence of Hurricane Intensity on Climate. *Nature,* 326:483–485.

Fujita, T. T., 1993. Wind Fields of Andrew, Omar, and Iniki, 1992. *Preprint from the 20th Conference on Hurricanes and Tropical Meteorology.* Boston: American Meteorological Society, 46–49.

Goldenberg, S., and L. Shapiro, 1996. Physical Mechanisms for the Association of El Niño and West African Rainfall with Atlantic Major Hurricane Activity. *J. Climate,* 9:1169–1187.

Gray, W. M., and W. M. Frank, 1993. Hypothesis for Hurricane Intensity Reduction from Carbon Black Seeding. *Preprint from the 20th Conference on Hurricanes and Tropical Meteorology.* Boston: American Meteorological Society, 305–308.

Gray, W. M., J. D. Sheaffer, and C. W. Landsea, 1996. Climate Trends Associated with Multi-Decadal Variability of Intense Atlantic Hurricane Activity. Chapter 2 in *Hurricanes, Climatic Change and Socioeconomical Impacts: A Current Perspective,* H. F. Diaz and R. S. Pulwarty, eds. Boulder, CO: Westview Press, 49 pp.

Gray, W. M., C. W. Landsea, P. W. Mielke Jr., and K. J. Berry, 1998. Extended Range Forecast of Atlantic Seasonal Hurricane Activity and U.S. Landfall Strike Probability for 1999. Department of Atmospheric Science Paper, Colorado State University. Also available at http://tropical.atmos.colostate.edu.

Hebert, P. J., J. D. Jarrell, and M. Mayfield. 1996. The Deadliest, Costliest, and Most Intense United States Hurricanes of This Century (and Other Frequently Requested Hurricane Facts). *NOAA Technical Memorandum NWS TPC-1.* Miami: National Oceanic and Atmospheric Administration, National Weather Service, National Hurricane Center, 30 pp.

Henderson-Sellers, A., H. Zhang, G. Berz, K. Emanuel, W. Gray, C. Landsea, G. Holland, J. Lighthill, S-L. Shieh, P. Webster, and K. McGuffie, 1998. Tropical Cyclones and Global Climate Change: A Post-IPCC Assessment. *Bull. Am. Meteor. Soc.,* 79:19–38.

Knaff, J., 1997. Implications of Summertime Sea Level Pressure Anomalies in the Tropical Atlantic Region. *Mon. Wea. Rev.,* 10:789–804.

Knaff, J. A., 1998. Predicting Summertime Caribbean Sea Level Pressure. *Wea. Forecasting*, 13:740–752.

Landsea, C., 1998. Frequently Asked Questions: Hurricanes, Typhoons, and Tropical Cyclones. Available at http://www.aoml.noaa.gov/hrd/tcfaq/.

Landsea, C. W., 1999. Climate Variability of Tropical Cyclones: Past, Present, and Future. In *Storms*. Edited by Roger Pielke Jr. and Roger Pielke Sr. United Kingdom: Routledge Press.

Landsea, C. W., N. Nicholls, W. M. Gray, and L. A. Avila, 1996. Downward Trends in the Frequency of Intense Atlantic Hurricanes During the Past Five Decades. *Geo. Res. Letters*, 23:1697–1700.

Leggett, J., ed., 1994. *The Climate Time Bomb*. Amsterdam: Greenpeace International.

LePore, F., 1996. Interview with the daily science radio show *Earth and Sky*. Transcript available from Earth and Sky, or at ftp://ftp.earthsky.com.

Nese, J. M., and L. M. Grenci, 1998. *A World of Weather: Fundamentals of Meteorology*. Kendall/Hunt Publishing Co., 539 pp.

Neumann, C. J., 1993. Global Overview. In *Global Guide to Tropical Cyclone Forecasting*. World Meteorological Organization Technical Document, WMO/TD No. 560, Tropical Cyclone Programme, Report No. TCP-31. Edited by G. Holland. Geneva, Switzerland, Chapter 1. Also available at http://www.bom.gov.au/bmrc.

Newsweek, 1996. The HOT ZONE: Hurricanes, Floods, and Blizzards: Blame Global Warming. January 21.

Newton, D. E., 1993. *Global Warming*. Santa Barbara, CA: ABC-CLIO, 183 pp.

Pielke, R. A., Jr. and C. W. Landsea, 1998. Normalized Atlantic Hurricane Damage: 1925–1995. *Wea. Forecasting*, 13:621–631.

Pielke, R. A., Jr. and R. A. Pielke Sr., 1997. *Hurricanes. Their Nature and Impacts on Society*. Chichester, England: John Wiley & Sons, 279 pp.

Posey, C., 1994. Hurricanes—Reaping the Whirlwind. *Omni*, 16:34–47.

Rosenfeld, J., 1997. Storm Surge! Hurricanes' Most Powerful and Deadly Force. *Weatherwise*, 50:18–24.

Ryan, B. F., I. G. Watterson, and J. L. Evans, 1992. Tropical Cyclone Frequencies Inferred from Gray's Yearly Genesis Parameter: Validation of GCM Tropical Climates. *Geo. Res. Letters*, 24, 1255–1258.

Simpson, R. H., 1974. The Hurricane Disaster Potential Scale. *Weatherwise*, 27:169, 186.

U.S. Department of Commerce, 1969. *Hurricane Camille, August 14–22, 1969 (Preliminary Report)*. Washington, DC: Environmental Science Services Administration, Weather Bureau, 58 pp.

U.S. Senate Bipartisan Task Force on Funding Disaster Relief, 1995. *Federal Disaster Assistance*. U.S. Senate Listing Number 104–4.

Willoughby, H. E., J. A. Clos, and M. G. Shoreibah, 1982. Concentric Eyewalls, Secondary Wind Maxima, and the Evolution of the Hurricane Vortex. *J. Atmos. Sci.*, 41:1169–1186.

Willoughby, H. E., D. P. Jorgensen, P. G. Black, and S. L. Rosenthal, 1985. Project STORMFURY: A Scientific Chronicle 1962–1983. *Bull. Am. Meteor. Soc.*, 66:505–514.

Willoughby, H. E., and R. A. Black, 1996. Hurricane Andrew in Florida: Dynamics of a Disaster. *Bull. Am. Meteor. Soc.*, 77:543–549.

Chronology 2

This chapter is divided into three parts. The first part chronologically lists significant U.S. landfalling hurricanes since 1900 with commentary about their impact. The second part chronologically lists scientific advances in hurricane research and forecasting. The last part discusses pre-1900 hurricanes that have altered history.

Table 2.1 lists in chronological order some significant U.S. landfalling hurricanes since 1900. The table also includes (1) the area or areas affected; (2) the Saffir-Simpson category based on the central pressure at landfall (should be treated with caution since central pressure does not correlate strongly with maximum sustained wind speed); (3) the number of U.S. fatalities associated with the storm (many of these figures should be considered estimates only, especially for the earlier storms); and (4) general comments on the storm history and the damage it caused. This section is an updated version of an out-of-print U.S. Department of Commerce NOAA technical report *Some Devastating North Atlantic Hurricanes of the Twentieth Century*. This report listed a table similar in format to Table 2.1 for hurricanes from 1900 to 1980; however, I have made significant changes to the comments as well as

TABLE 2.1
Chronology of Landfalling U.S. Storms since 1900

Dates of Hurricane	Areas Most Affected	Saffir-Simpson Category at Landfall	Deaths (U.S. only)	Comments and Damage
1900, August 27–September 15 Galveston Hurricane	Texas	Category 4	At least 6,000 on island and 2,000 more inland	Principle damage and most loss of life is caused by 8- to 15-foot storm surge that inundates Galveston Island. Flying debris causes some deaths. Damage is widespread along Texas coast. Of the island's roughly 16,000 inhabitants, 6,000 are killed, and the others are bruised and battered. This is still the worst U.S. natural disaster in terms of fatalities. After the storm, it is discovered that most survivors were located behind concentrated, elongated areas of debris that acted as a break against the full force of the storm surge, and the notion of a seawall is conceived to help prevent future disasters. A six-mile-wide seawall is completed in 1905, and the elevation of the entire city is raised 8 feet, to a height of 17 feet.
1906, September 19–29	Florida panhandle to Mississippi	Category 3	134	Destructive winds and unprecedented tides accompany the storm. At Pensacola, Florida, the tide is 10 feet above normal, and the storm surge contributes to severe property damage.
1909, July 13–22	Texas	Category 3	41	Hurricane passes directly over Velasco, Texas. Town is in the calm of the eye for 45 minutes, followed by devastating winds that destroy half the town.
1909, September 14–21	Louisiana and Mississippi	Category 4	350	Hurricane center passes 50 to 75 miles west of New Orleans. Wide extent of Louisiana coast is inundated.
1915, August 5–25	Texas and Louisiana	Category 4	275	Despite the seawall and raised island, a storm surge of 12 feet above normal floods the Galveston business district to a depth of 6 feet. Nevertheless, the seawall's value is evident because only 8 are killed in the city. Along adjacent coastal areas, Galveston Bay, and the parts of Galveston Island without the seawall, destruction is massive and 267 are killed.
1915, September 22–October 1	Louisiana	Category 4	275	Ninety percent of buildings are destroyed over a large area of Louisiana south of New Orleans and along Lake Pontchartrain. Many casualties are caused by people remaining in low-lying areas despite warnings.
1916, June 29–July 10	Mississippi to northern Florida	Category 3	7	This hurricane arrives unexpectedly, leaving shoppers and businessmen stranded in downtown Mobile hotels. Eight feet of water inundates the lobby of the Grand Hotel at Point Clear. This hurricane is very destructive along the coast from Mobile to Pensacola.
1919, September 2–15	Florida, Louisiana, and Texas	Category 4 in Florida Keys and	287 on land, 500 on ships at sea	Hurricane is severe both in Florida and in Texas. The slow-moving storm reaches an intensity of 927 mb 65 miles west of Key West. Ten vessels are lost at sea, accounting for 500 casualties. The hurricane

	Location	Deaths	Category	Description
			Category 4 in Texas	moves slowly westward and travels inland south of Corpus Christi, inundating the region with 16 feet of water and killing 287 people. Much of the Texas coast is devastated by high water.
1926, September 11–22 Great Florida Storm	Florida and Alabama	243	Category 4 in southeast Florida; Category 3 at Alabama-Florida border	Hurricane is very severe in Miami area and from Pensacola into southern Alabama. Many go outside during the eye's calm interval and are killed when the strong winds return. The storm "bends" Miami's newest skyscraper, the Myer-Kiser Bank Building, 15 degrees, forcing the building to be torn down. The surge drives a steamer into the middle of Miami. Miami's economy, which had been expecting a development boom, suffers because of the destruction and lost investment.
1928, September 6–20 Okeechobee Hurricane	Southern Florida	1,836	Category 4	This hurricane kills 1,836 people and injures 1,849 on the shores of Lake Okeechobee, the second largest freshwater lake in the United States. It is the second deadliest natural event in U.S. history. Most of the deaths are by drowning, but some are due to snakebites as people climb into trees to escape the flood waters, only to find hordes of venomous water moccasins also seeking shelter. Although the lake is actually inland, the hurricane forces a storm surge that breaks the eastern earthen dike on the southern end of Lake Okeechobee. This calamity occurs within a few miles of a large city and a world-famous resort, yet so isolated is the location that no one knows what happened for three days afterward. A similar situation to another hurricane occurred two years earlier and killed about 200 at the Okeechobee town of Moore Haven. As a result of these disasters, a levee (called the Herbert Hoover Dike in honor of the president who supported its construction) that compares in size to the Great Wall of China is built to prevent further disasters.
1932, August 11–14	Texas	40	Category 4	Landfall occurs near Freeport. Hurricane is small in diameter; winds damage rice and some cotton near the coast, but the accompanying rains are beneficial to the drier interior areas.
1933, August 17–26	North Carolina, Virginia, and Maryland	47	Category 3	Storm makes landfall at Nags Head on the outer banks and travels west of Norfolk, Virginia; Washington, D.C.; Baltimore, Maryland; and Atlantic City, New Jersey. It is the only storm to strike the Chesapeake-Potomac area in the twentieth century. Heavy damage occurs in northeastern North Carolina and resorts on Maryland, Delaware, and New Jersey coasts; downtown Norfolk floods; and crop damage is heavy in Maryland and Virginia.
1933, August 28–September 5	Texas	40	Category 3	Heavy property damage from Corpus Christi to northeastern Mexico; citrus crop almost completely destroyed.
1933, August 31–September 7	Florida	2	Category 3	Much property damage on the coast along Vero Beach; property damage inland is minor; citrus loss is nearly complete near the coast.

(continues)

TABLE 2.1
(continued)

Dates of Hurricane	Areas Most Affected	Saffir-Simpson Category at Landfall	Deaths (U.S. only)	Comments and Damage
1933, September 8–21	North Carolina	Category 3	21	Heavy damage from Carteret County to Virginia state line; high storm surge in Pamlico and Albermarle Sounds.
1934, June 4–21	Louisiana	Category 3	6	Property and crop damage is moderate.
1934, July 21–25	Florida and Texas	Category 2	11	Heavy rains severely damage the Texas cotton crop.
1935, August 29– September 10 Labor Day Storm	Florida Keys and northwest Florida	Category 5 in Florida Keys; Category 2 in northwest Florida	409	Barometer reading of 26.35 inches (892 mb) on Long Key is the lowest on record in the Western Hemisphere until Hurricane Gilbert (1988). Peak winds are estimated 150–200 mph on some keys with an 18-foot storm surge, making it still the strongest U.S. landfalling hurricane (Gilbert never hit the United States). The Labor Day storm is an example of a "midget hurricane" that is very small in diameter and a striking example that hurricane intensity has little to do with size. It clears the entire landscape of every tree and every building on Matecumbe Key. The dead include residents and 259 World War I unemployed veterans building a road as part of the Work Projects Administration. A train is sent to evacuate them, but when it tries to return north the storm surge has washed out the track. Shortly thereafter, the train is washed off the track and everyone is killed. Ernest Hemingway visits the site and writes about the destruction in a scathing article called, "Who Killed the Vets" for *New Masses* magazine.
1935, October 30– November 8 Yankee Storm	Southern Florida	Category 2	5	This storm forms near Bermuda and moves north to south along the East Coast, hitting the Miami area. Because of its unusual track, locals call it the "Yankee Storm" to counter hecklers in the Northern States who have started calling these storms "Florida hurricanes" because of the recent frequency of Florida landfalls. It is quite small — destructive winds cover only a narrow path, and because it hits an uninhabited area, casualties are small.
1938, September 10–22 New England Hurricane	Long Island, New York and southern New England	Category 3	600	This hurricane is sometimes called the "Long Island Express" because of its rapid northward motion (up to 56 mph) toward the Long Island coast. This storm surprises New England residents who are unaccustomed to hurricanes (the last hurricane had hit New England 70 years before). Because of the hurricane's first motion and because the Weather Bureau thought it would recurve offshore and not hit the United States, there is little preparation time available. Even worse, the storm hits at the equinox when tides are usually their highest. Ten-foot waves atop a 14- to 18-foot storm surge kills 600 people and causes immense

Date	Location	Category	Deaths	Description
				property damage. Barrier islands are swept so bare that rescue workers use phone company charts to determine where houses once stood. Sustained hurricane winds penetrate far inland, causing extensive damage to roofs, trees, and crops. In Connecticut, downed power lines result in catastrophic fires to sections of New London and Mystic. Rainfall results in severe flooding across sections of Massachusetts and Connecticut. This storm holds the record for the most property damage in the United States until 1954.
1940, August 5–15	Georgia and the Carolinas	Category 2	50	Heavy flooding in the southeastern states as far inland as Tennessee from hurricane-induced rains. Thirty deaths are due to floods.
1941, September 16–25	Texas	Category 3	4	Very heavy crop damage. Most low, exposed places are evacuated in response to good warnings, resulting in low casualties.
1943, July 25–29 The Surprise Hurricane	Texas	Category 2	19	Because of a "dare" from a fellow officer, Army Air Corps Col. Joseph P. Duckworth flies the first intentional flight into this hurricane. This hurricane is also noteworthy because it hits with little warning, killing 19 people, injuring hundreds, and causing significant property damage along Galveston Bay. Because of World War II censorship, a forecaster's major source of information—ship reports—are unavailable so German U-boats will not get weather information. Therefore, the Weather Bureau, who thinks they are dealing with a tropical storm, do not know it is a Category 2 hurricane until it has made landfall. Forecast advisories are often several hours delayed since they require security clearance, even after landfall. Barometer readings are also restricted from the general public as the hurricane approaches. All of this contributes to a confused and uninformed public. It could have been much worse if Galveston had no seawall or if the storm had hit the island itself (it made landfall on the Bolivar peninsula to the east). After this, never again are forecast advisories censored from the general public.
1944, September 9–16 The Great American Hurricane	North Carolina to New England	Category 3	344	This storm receives its name because it travels up the Atlantic coast from North Carolina to the Northeast, making landfall in Rhode Island. Heavy damage occurs, but it is one-third as great as the 1938 hurricane. U.S. fatalities are reduced (46 killed) because of better preparation and because the stronger portion of storms remains offshore of New England. However, the U.S. Navy and the Coast Guard lose 5 ships offshore, and the storm kills 298 servicemen.
1944, October 12–23	Southwest Florida	Category 3	18	Warnings and evacuation prevents heavier casualties.
1945, September 11–20	Southeast Florida	Category 3	4	Damage very heavy in Dade County (Miami). Evacuation of exposed locations prevents heavy loss of life.
1947, September 4–21	Southeast Florida, Louisiana, and	Category 4 in Florida;	51	This severe hurricane crosses over Florida and hits Louisiana and Mississippi. The center of this very large and intense storm hits Hillsboro Light, Florida, on September 17 with gusts to 155 mph. After leaving

(continues)

TABLE 2.1
(continued)

Dates of Hurricane	Areas Most Affected	Saffir-Simpson Category at Landfall	Deaths (U.S. only)	Comments and Damage
	Mississippi	Category 3 in Louisiana and Mississippi		Florida, the huge hurricane takes a northwesterly course over the Gulf of Mexico and onto the Mississippi and Louisiana coasts. Tides rise to 12 feet at Biloxi, Bay St. Louis, and Gulfport, Mississippi. The eye of the storm passes directly over New Orleans and is estimated at 12 miles in diameter. A total of 51 lives are lost—17 in Florida, 12 in Louisiana, and 22 in Mississippi. Damage is heavy in all three states.
1947, October 9–16	Southern Florida, Georgia, and South Carolina	Category 1 in South Florida; Category 2 in Georgia and South Carolina	1	The northward-moving hurricane hits southern Florida, moves offshore east of Florida, then turns west toward the Georgia–South Carolina border. Heavy rains in Florida climax a very wet season. Heavy damage also occurs in the Savannah area from wind and along the Georgia–South Carolina coast from high tides.
1948, September 18–25	Southern Florida	Category 3	3	Heavy damage occurs. Many reports of calm winds from widely separated simultaneous points suggest a large eye or double eyewall.
1948, October 3–15	Southern Florida	Category 2	0	The damage is not as great as could be expected because much of the area has already experienced damage from the September storm.
1949, August 23–31	Florida to the Carolinas	Category 3	2	Storm center passes over Lake Okeechobee. Levees built since 1928 prevent overflow and casualties.
1949, September 27–October 6	Texas	Category 2	2	Most of damage is to crops.
1950, September 1–9 Easy	Northwest Florida	Category 3	2	Also called the "Cedar Key Hurricane," this storm's path exhibits an unusual double loop near landfall near Cedar Key. The coast from Sarasota northward suffers extensive wind, storm surge, and flood damage. Nearly 39 inches of rain fall on Yankeetown in 24 hours. The coastal area inland from Yankeetown to Tampa is flooded for several weeks.
1950, October 13–19 King	Southeast Florida	Category 3	4	This is a small, violent storm that passes directly over Miami, then up the entire inland Florida peninsula.

Date/Name	Location	Category	Deaths	Description
1954, August 25–31 Carol	North Carolina to New England	Category 2 in North Carolina; Category 3 in New York, Connecticut, and Rhode Island	60	Property losses are the greatest from any single storm up to this date. Extremely high tides flood many low-lying regions in New England.
1954, September 2–14 Edna	New Jersey to New England	Category 3	21	Heavy damage occurs again in New England.
1954, October 5–18 Hazel	South Carolina to New York	Category 4 in South and North Carolina	95	Heavy damage occurs in exposed North Carolina shore areas due to a high storm surge superimposed on the highest ocean tide of the year. The most destructive storm in North Carolina's history, Hazel produces wind gusts of 150 mph. Hazel retains its intensity inland as it joins with another storm system to devastate inland communities from Virginia to Ontario, Canada. Washington, D.C., experiences its strongest winds on record. Because of Hurricanes Carol, Edna, and Hazel, as well as other previous East Coast hurricanes, Congress increases funding for hurricane research and the National Hurricane Research Laboratory is established.
1955, August 3–14 Connie	North Carolina	Category 3	25	Heavy rainfall (between 6 and 12 inches) falls between North Carolina and New England. This rainfall saturates the ground and fills the streams, thus setting the stage for Diane.
1955, August 7–21 Diane	North Carolina to New England	Category 1	184	Due to high watersheds and saturated ground from Connie, Diane's heavy rainfall causes severe floods throughout the northeast United States. Damage exceeds any prior storm on record, earning itself the nickname the "Billion-Dollar Hurricane" — the first storm to reach this damage threshold. (Ironically, later estimates show it was less, but damage was truly extensive.)
1955, September 10–23 Ione	North Carolina	Category 3	7	Ione is the third hurricane in eastern North Carolina within 5 weeks and the fourth in 11 months.
1956, September 21–30 Flossy	Louisiana to northern Florida	Category 2 in Louisiana; Category 1 in Florida	15	Damage occurs over an area from New Orleans to the mouth of the Mississippi, and eastward to western Florida. Flossy completely submerges Grand Isle and causes extensive coastal erosion in Louisiana marshland. Hundreds lose homes; cattle are drowned; and citrus, sugar, and pecan crops are heavily damaged. The eastern section of New Orleans' seawall is overtopped with inland flooding.

(continues)

TABLE 2.1
(continued)

Dates of Hurricane	Areas Most Affected	Saffir-Simpson Category at Landfall	Deaths (U.S. only)	Comments and Damage
				Heavy rain occurs in Louisiana and Alabama. Evacuations are a success; deaths are mainly due to plane and car accidents.
1957, June 25–28 Audrey	Texas and Louisiana	Category 4	526	Hurricane Audrey makes landfall near the Texas-Louisiana border on June 27 with devastating effects and is the most destructive hurricane to strike southwest Louisiana in history. Its central pressure deepens considerably in the last five hours before landfall. There are many deaths as the result of a storm surge in excess of 12 feet, which inundates the flat coast of Louisiana as far as 25 miles inland in some places. Many homes are destroyed, and offshore oil installations are heavily damaged. Texas also experiences a high storm surge, with 450 people injured.
1959, September 20–October 2 Gracie	South Carolina to Virginia	Category 3	22	Storm passes inland on the South Carolina coast near Beaufort with wind gusts near 138 mph, causing the most severe property and crop damage in history in Beaufort. Twelve deaths are associated with tornadoes in Virginia; car accidents, falling trees, and live wires cause 10 deaths.
1960, August 29–September 13 Donna	Florida to New England	Category 4 in southwest Florida and Keys; Category 3 in North Carolina and New York; Category 2 in Connecticut and Rhode Island	50	At the time, this hurricane produces a record amount of damage in Florida. It makes landfall as a hurricane in the Florida Keys, southwest Florida, North Carolina, New York, and Connecticut—the first time that has happened since records were first taken. Winds are estimated near 140 mph, with gusts 175–180 mph on central Florida Keys.
1961, September 3–15 Carla	Central Texas	Category 4	46	Largest and most intense Gulf Coast hurricane in decades, Carla impacts Corpus Christi to the Louisiana border with hurricane-force winds. The eye passes over Port O'Connor, totally devastating the town. Severe damage occurs along a wide expanse of the Texas coast from unusually prolonged winds, high tides, and flooding. In addition, Galveston is seriously affected by a tornado. The region's rice crop is ruined, and entire herds of cattle drown.

Date	Name	Location	Category		Description
1964, August 20– September 5	Cleo	Southern Florida and eastern Virginia	Category 2	0	First hurricane in Miami area since 1950. Moderate wind damage is extensive along Florida's lower east coast. Record rainfall and widespread floods from Hampton Roads area southward in Virginia. Tornadoes occur in southeast Florida and the Carolinas.
1964, August 28– September 16	Dora	Northeastern Florida and southern Georgia	Category 2	5	Dora is the first storm of full hurricane force on record to move inland from the east over extreme northeastern Florida, causing extensive wind damage, inland flooding, and crop damage in Florida; Georgia also suffers damage. Storm slows considerably before landfall, perhaps allowing time for residents to evacuate who are unaccustomed to hurricanes. The persistent onshore winds produce an estimated storm surge of 12 feet that severely erodes beaches, inundates several coastal communities, and washes out beach roads.
1964, September 28– October 5	Hilda	Louisiana	Category 3	38	Damage from wind, rain, and tornadoes is extensive in southeast Louisiana. On the morning of October 3, several tornadoes occur in southeastern Louisiana in prehurricane squall lines. One tornado at Larose kills 22 and injures 200. Three other tornadoes cause much damage in the New Orleans metropolitan area but no deaths. Hurricane Hilda's highest reported wind is an estimated 135 mph at Franklin. An almost complete evacuation of the entire Louisiana coast accounts for the low death toll of 38, including the tornado fatalities. Offshore oil facilities also are damaged.
1965, August 27– September 12	Betsy	Southern Florida and Louisiana	Category 3 in Florida and Louisiana	75	Betsy moves south through the Bahamas, then west over the Florida Keys. Damage from winds, high tides, and wave action is confined to an area from Ft. Lauderdale southward. Flooding over the upper Keys is extensive. Betsy turns to the northwest upon entering the Gulf of Mexico, with its eye arriving at Grand Isle, Louisiana, the evening of September 9. The eye is 40 miles in diameter on the Louisiana coast. Great devastation is caused by high water on the central Gulf Coast from the point where the center makes landfall to Mobile, Alabama. A 10-foot storm surge in New Orleans produces the city's worst flooding in decades. Following the storm, the levee surrounding the city is raised to 12 feet. Evacuation advice prompts 300,000 people in Louisiana to seek safe shelter. However, 58 people lose their lives because of widespread wind gusts in excess of 100 mph and floods in Louisiana. Highest sustained winds of 136 mph are recorded at Port Sulphur, Louisiana, with gusts to 160 mph reported along the Gulf Coast. There are four deaths in Florida; other lives are lost in the adjacent waters of the Gulf and the Atlantic. The total of 75 deaths in Betsy is the greatest loss of life along the Gulf Coast since Audrey in 1957.
1967, September 5–22	Beulah	Southern Texas	Category 3	15	Beulah makes landfall between Brownsville and the mouth of the Rio Grande about daybreak on September 20. Beulah produces a storm surge of at least 18 feet, which creates 31 new cuts on Padre Island. This slow-moving, erratic storm dumps torrential rains of 10 inches or more on a vast area of

(continues)

TABLE 2.1
(continued)

Dates of Hurricane	Areas Most Affected	Saffir-Simpson Category at Landfall	Deaths (U.S. only)	Comments and Damage
				south Texas from San Antonio southwestward to the Rio Grande and southeastward to the Gulf of Mexico (40,000 square miles), producing major floods that kill 10 people and cause most of the damage. In addition, a record number of tornadoes—155—occur, which kill 5 people.
1969, August 14–22 Camille	Mississippi, Louisiana, Alabama, Virginia, and West Virginia	Category 5	262	Hurricane Camille, with sustained winds of at least 180 mph, produces a storm surge of 23 feet in Pass Christian, Mississippi, which kills 137 people in Mississippi and 9 in Louisiana, including 21 of 23 people who ignore evacuation warnings and stay for a "hurricane party." A survivor of the hurricane party, who clung to floating debris, traveled 10 miles or more in the storm surge. Floods from the storm surge penetrate 8 miles inland in the Waveland–Bay St. Louis region, and in river estuaries the flood extends 20 to 30 miles upstream. Camille's storm surge also pushes water over both levees near the mouth of the Mississippi River in Louisiana, removing nearly all traces of civilization, as a 1969 U.S. Department of Commerce report states. Total devastation on the immediate coastline and severe damage further inland occurs. Camille also dumps 30 inches of rain in 6 hours in the Blue Ridge Mountains, triggering flash floods and mud slides that kill 114 people in Virginia and 2 in West Virginia.
1970, July 23–August 5 Celia	Texas	Category 3	11	Celia intensifies rapidly in the 15 hours before it crosses the coast north of Corpus Christi. As it moves over land, spectacular damage occurs from high-energy winds of short duration (also called downbursts or microbursts) that exceed the prevailing sustained wind by a factor of two or three. The extreme winds rake across the residential and business areas in less than half an hour. It is estimated that winds reach as high as 160 mph for several seconds. During those disastrous seconds, incredible damage occurs at the airport and an adjacent mobile home park that is completely demolished. Fortunately, only 11 die in the Corpus Christi area due to the state of preparedness by its disaster prevention agencies. This is the costliest hurricane to strike the Texas coast up to that time.
1972, June 14–23 Agnes	Florida to New York	Category 1 in Florida and in New York	122	Agnes is one of the largest June hurricanes in history, with a circulation over 1,000 miles in diameter. After landfall, a weakening Agnes interacts with another storm system so that it has two centers, and it actually regains tropical storm strength over land in North Carolina. The main part of the storm moves offshore and eventually makes landfall on the western tip of Long Island. Agnes produces devastating floods from North Carolina to New York, with many record river crests and over 100 deaths. Agnes is

one of the costliest natural disasters in U.S. history. Agnes also produces 15 tornadoes in Florida and 2 in Georgia that cause localized damage.

Date/Name	Category	Location		Description
1974, August 29–September 10, Carmen	Category 3	Southern Louisiana	1	Carmen moves inland just east of Vermilion Bay with wind gusts to 120 mph. Damage occurs primarily to sugar cane crop, offshore installations, and the shipping industry.
1975, September 13–24, Eloise	Category 3	Florida panhandle and eastern Alabama	4	Eloise makes landfall about midway between Fort Walton Beach and Panama City, Florida, early on September 23. Winds are estimated at about 100 mph with a storm surge of 12–16 feet. The storm surge and wind cause extensive damage to structures along the 25-mile beach strip from Fort Walton Beach to Panama City. High winds destroy property and crops over eastern Alabama. Flooding and miscellaneous damage from heavy rains occurs over northeastern United States.
1979, August 25–September 7, David	Category 2 in Florida, Georgia, and South Carolina	Eastern Florida through middle Atlantic states	5	Damage is not great in any one area, but the cumulative total caused by winds, storm surge, floods, and tornadoes is large because of the wide area affected. Numerous tornadoes occur along inland path.
1979, August 29–September 14, Frederic	Category 3	Southern Alabama, southern Mississippi, and Florida panhandle	5	The eye passes over Dauphin Island, Alabama, on September 13 with sustained winds of 120 mph and gusts to 145 mph. This storm causes extensive damage with its 15-foot storm surge near the mouth of Mobile Bay. Major wind damage occurs in southern Alabama, southern Mississippi, and the extreme western Florida panhandle, making it one of the most costly Gulf Coast hurricanes in history. Heavy rains in Ohio and New York result from the interaction of a weakening Frederic and a cold front, which produces record floods and major damage in this region.
1980, August 1–11, Allen	Category 3	South Texas	2	Originally a very powerful Category 5 hurricane, Allen slows down and weakens to a Category 3 hurricane before making landfall just north of Brownsville, hitting a largely unpopulated area. A storm surge up to 12 feet along Padre Island causes numerous cuts.
1983, August 15–21, Alicia	Category 3	North Texas	21	An unusual hurricane, Alicia forms in an environment that appears unfavorable for development. Local officials in Galveston decide against complete evacuation of coastal areas. When Alicia intensifies significantly the last 18 hours before landfall, it is too late to evacuate these areas. Only the presence of Galveston's seawall prevents a large loss of life. Alicia makes landfall 25 miles southwest of Galveston, then passes directly over Houston, blowing out skyscraper windows and causing extensive damage.

(continues)

TABLE 2.1
(continued)

Dates of Hurricane	Areas Most Affected	Saffir-Simpson Category at Landfall	Deaths (U.S. only)	Comments and Damage
				Alicia also produces 26 tornadoes. Even though Alicia is only a small- to medium-sized storm, it leaves a legacy as the costliest hurricane in Texas history.
1985, August 28—September 4 Elena	Louisiana to west Florida	Category 3	4	Elena is best remembered for its bizarre track, which confounded forecasters. On August 29, Elena is following a northwest path that would bring landfall somewhere between Louisiana and Florida. But on August 30, the steering currents change and Elena moves eastward toward the west coast of Florida. Within a couple hundred miles of Cedar Keys, Florida, Elena stalls for two days before turning back to the west-northwest, finally making landfall near Biloxi, Mississippi. Nearly a million people are evacuated, with a large number in the middle Gulf Coast being asked to evacuate twice within a three-day period. This is the largest number of people ever evacuated for a hurricane to date. Elena inflicts a devastating blow to the oyster industry in Apalachicola, Florida, as well as extensive beach erosion and damage from the mouth of the Mississippi River to Fort Myers, Florida.
1985, September 16—October 2 Gloria	North Carolina to New York	Category 3 in North Carolina; Category 2 in Long Island	8	Gloria reaches a minimum pressure of 919 mb about 900 miles southeast of Cape Hatteras, which is the lowest pressure ever measured by reconnaissance aircraft over the East Coast Atlantic region. Gloria receives much attention from the media because of the threat it poses to the northeastern United States. Fortunately, Gloria weakens as it moves northward, somewhat minimizing casualties and damage. It crosses over the Outer Banks of North Carolina on September 27, passes just offshore of the mid-Atlantic states, and makes landfall over western Long Island, New York, ten hours later. The strongest winds of Gloria generally remain offshore throughout this period, and Gloria hits New York during low tide, thus minimizing the storm surge. Downed trees cause extensive power outages for hundreds of thousands in the Long Island and New England area. Beach erosion is severe from North Carolina to New York, and North Carolina experiences coastal flooding.
1985, October 26—November 1 Juan	North Texas to Florida	Category 1 in Louisiana; tropical storm in Florida	12	Juan is a weak but persistent storm that raises much havoc on the northern Gulf Coast, with three landfalls resulting from multiple loops in its track. Even before Juan becomes a tropical storm, the tropical disturbance creates a broad zone of gale-force winds, which makes evacuation of offshore rigs very difficult as Juan is developing. Nine deaths are caused by toppled oil rigs or sunken boats rescuing oil workers. Juan makes its first loop just off the Louisiana coast on October 28 and makes landfall near Morgan City on October 29. The following day Juan makes a second loop around Lafayette and emerges over Vermilion Bay, where it becomes better organized. It moves over the mouth of the

Date and Name	Category	Location	Deaths	Description
				Mississippi River near Burwood, Louisiana, then finally makes landfall just west of Pensacola, Florida. Because of Juan's slow, erratic movement, it creates persistent onshore flow and excessive rain, causing extensive damage from the upper Texas coast to the extreme western Florida panhandle. The damage affects offshore oil rigs, crops, and livestock when high water overflows levees in south Louisiana and areas flooded around Lake Pontchartrain.
1988, September 8–19 Gilbert	Did not hit United States; hit Mexico as Category 3	South Texas	3	Gilbert is noteworthy because it produces the lowest surface pressure ever recorded in the Western Hemisphere of 26.33 inches (888 mb) while over open ocean. This storm causes much damage and a number of fatalities in Mexico, Jamaica, Haiti, and Central America. U.S. damage is confined to 29 tornadoes in south Texas from the remnants of Gilbert.
1989, September 10–22 Hugo	Category 4	South Carolina and North Carolina	21	Hugo is the most powerful hurricane to strike the United States since Camille in 1969, with sustained winds of 130 mph or more and extreme storm surges of 20 feet in some places. It is also the costliest hurricane until Andrew in 1992. Hugo makes landfall near Charleston at Sullivans Island on September 22, devastating waterfront properties and beaching (or destroying) an incredible number of watercraft. Hugo's fairly swift motion carries the hurricane well inland before it can significantly weaken. Almost two-thirds of South Carolina receives wind damage, as does much of central North Carolina along a wide north-south band through the Charlotte area. Falling trees and limbs do much of the damage; otherwise, roof damage is common. The timber industry is wiped out, and power outages are widespread.
1991, August 16–29 Bob	Category 2	New England	17	Bob makes landfall on Rhode Island, then passes between Boston and Sciuate, Massachusetts. Although Bob is a weakening hurricane, it still causes widespread damage. Dozens of cottages and boats on Narragansett Bay are damaged or destroyed by a seven-foot storm surge. Heavy rain of four to seven inches in northwest Rhode Island causes considerable urban flooding. Other portions of New England also experience damage from the storm surge, downed trees, flooding from heavy rain, and tornadoes. Power outages are widespread. Apple crops in southern Maine suffer extensive damage.
1992, August 16–28 Andrew	Category 4 in Florida; Category 3 in Louisiana	South Florida and Louisiana	26	Andrew is a small but destructive hurricane that devastates south Florida and south-central Louisiana. Damage resulting from the hurricane is extreme, making Andrew the most expensive natural disaster in U.S. history. Injuries are numerous, and 250,000 people are left homeless. A record 17-foot storm surge is recorded near the Burger King headquarters south of Miami, causing much localized damage. However, much inland damage is caused by winds in Andrew's eyewall, which levels homes and businesses. Damage is extreme from the Kendall district southward through Home-

(continues)

TABLE 2.1
(continued)

Dates of Hurricane	Areas Most Affected	Saffir-Simpson Category at Landfall	Deaths (U.S. only)	Comments and Damage
				stead and Florida City to near Key Largo, with government officials saying this area looks like a "war zone." So expensive is the damage that eight insurance companies cannot cover the claims and are driven out of business. Central Louisiana also suffers significant damage to its offshore platforms, barrier islands, and coastal properties due to winds and an 8-foot storm surge. One strong tornado causes damage in Laplace and Reserve, Louisiana. An upwelling of bottom sediment kills 187 million fish in the Atchafalaya Basin, tremendously affecting the Louisiana fishing industry. The sugar and soybean crops are devastated.
1992, September 5–13 Iniki	Hawaii	Category 4	6	The south side of the Kauai island around Poipu receives the most damage, but the northern shores of the island are equally ravaged. 10,000 homes are severely damaged, making upward of a third of the population temporarily homeless. Nearly all of the island's 70 hotels suffer damage. 7,000 of the island's 8,202 hotel, condo, and bed and breakfast rooms are shut down after the hurricane. Electricity and telephone service is not restored to many areas for several months. The hurricane cripples the island's tourist industry, which accounts for 45 percent of the island's economy. The high winds and surf also annihilate the island's entire crop of macadamia nuts and sugarcane, which are the island's other two large industries. Newly planted coffee trees and guava crops are also destroyed. It takes years for Kauai's tourism industry to rebuild and recover.
1993, August 22–September 6 Emily	North Carolina	Category 3	3	Emily's western eyewall passes over the North Carolina Outer Banks. Damage is confined mainly to Hatteras Island.
1995, July 31–August 6 Erin	East Florida and Florida panhandle	Category 1	6	Erin comes ashore near Vero Beach, causing widespread wind damage mostly in Brevard County. Some flooding occurs in the Melbourne and Palm Bay areas. After reemerging over the Gulf of Mexico, it makes landfall near Pensacola, causing significant structural damage in Navarre Beach and Pensacola Beach. Further inland, Erin knocks down trees and power lines. Crops suffer damage, with about half the cotton crop and a quarter of the pecan crop damaged. This is Pensacola's first direct hit by a hurricane since 1926. Remarkably, it is hit by Opal two months later.

Date/Name	Location	Category		Description
1995, September 27–October 5 Opal	Florida, Alabama, Georgia, and the Carolinas	Category 3	9	Opal makes landfall near Pensacola Beach as a minimal Category 3 hurricane, causing storm surge damage and erosion from southeastern Mobile Bay and Gulf Shores, Alabama, eastward through the Florida panhandle to Cedar Key, Florida. The structural damage is most extensive in the Fort Walton Beach area. Strong winds spread damage well inland. Opal downs numerous trees, knocking out power to 2 million people in Florida, Alabama, Georgia, and the Carolinas. The combination of Opal and a front results in heavy rain along the inland path of the hurricane. Opal also generates some concern the day before landfall. During the night, Opal unexpectedly intensifies explosively to a Category 4 with winds of 150 mph. Even worse, Opal begins accelerating toward Florida. Because most people are sleeping, it is difficult to alert the public about the new, unanticipated danger. Many people near Pensacola Beach awake the next morning with a near-Category 5 hurricane just offshore. The last-minute evacuation procedures clog the roads and Interstate 10. Fortunately, Opal begins to weaken, and coastal evacuations are completed in time.
1996, August 23–September 8 Fran	The Carolinas, Virginia, West Virginia, Maryland, and Pennsylvania	Category 3	34	Fran makes landfall over Cape Fear, North Carolina, but hurricane force winds extend over a wide area. A storm surge as high as 12 feet destroys or seriously damages numerous beachfront houses on the North Carolina coast. Most of the deaths are caused by flash flooding in the Carolinas, Virginia, West Virginia, Maryland, and Pennsylvania.
1998, September 21–30 Georges	South Florida, Louisiana, and Mississippi	Category 2 in Florida and Mississippi	4	After leaving a trail of destruction with 500 dead in the Caribbean region, Georges strikes the Florida Keys, leaving more than 900 homes with minor damage, 500 with major damage, and more than 150 homes completely destroyed, including 75 houseboats on so-called Houseboat Row. Georges makes its final landfall just east of Biloxi, Mississippi, causing a six-to-eight-foot storm surge east of the eyewall all the way to Mobile Bay, Alabama, where parts of downtown Mobile are flooded. Pasagoula, Mississippi, is particularly hard hit by a combination of the storm surge and flooding on the Pasagoula River. Storm surges above seven feet also occur in Lake Pontchartrain in Louisiana, destroying or damaging a number of fishing camps. Approximately 200 residents in the Florida panhandle must be rescued by the Coast Guard because of flooding in that region, and a portion of Interstate 10 near the Alabama-Florida state line is destroyed or washed over.
1999, August, 18–23 Bret	South Texas	Category 3	0	Bret is a potentially devastating storm but makes landfall over sparsely populated Kenedy County. Corpus Christi, 40 miles to the north, and Brownsville, 50 miles to the south, are spared. Preliminary estimates indicate most damage is to ranchlands.

(continues)

TABLE 2.1
(continued)

Dates of Hurricane	Areas Most Affected	Saffir-Simpson Category at Landfall	Deaths (U.S. only)	Comments and Damage
1999, August 24–September 5 Dennis	North Carolina, Virginia	Tropical Storm	0	Dennis threatens the Southeastern U.S. as a Category 2 storm, but turns sharply away from the coast as it nears landfall. Its legacy is its persistent nature—Dennis loops back, then stalls for a week before making landfall. The result is days of rain, gusty winds, and high surf.
1999, September 7–17 Floyd	North Carolina to New York	Category 2	79	Floyd is one of the most destructive and deadly hurricanes of the decade. Because of its size and intensity, the largest evacuation in U.S. history occurs as coastal residents from south Florida to North Carolina leave their homes. Floyd eventually hits Cape Fear, North Carolina as a Category 2 storm. Coastal damage due to wind and storm surge is relatively minor, but Floyd dumps 20 inches of rain in the region. 30,000 homes are flooded, dozens in eastern North Carolina drown.

added the Saffir-Simpson category at landfall. Also added are significant hurricanes since 1980 based on the National Hurricane Center's annual hurricane summaries published in the journal *Monthly Weather Review* (see Chapter 6 for a list). Table 2.1 does not include many landfalling tropical storms and weaker hurricanes. This omission does not imply that tropical storms should not be taken seriously, because they can still cause serious flooding (e.g., Tropical Storm Claudette in 1979 dumped 42 inches of rain in one day in Alvin, Texas). A detailed table of all U.S. landfalling hurricanes, as well as yearly storm tracks, is contained in Neumann et al. (1993).

Chronology of Weather Advances Related to the Study and Forecasting of Hurricanes

1687   Isaac Newton publishes *Principia,* the foundation of modern mathematical theory for all the physical sciences, including meteorology.

1743   Benjamin Franklin discovers that hurricanes do not necessarily move in the same direction as the surface wind direction. The cloud cover from a hurricane in 1743 blocks the eclipse Franklin wishes to see one night in Philadelphia. Later, he writes his brother in Boston about his disappointment and is surprised to find out the eclipse was visible in Boston, but soon afterward a violent storm hit. Franklin notes the difference in time between the onset of the storm in Philadelphia and Boston and reasons it must have been the same storm traveling from southwest to northeast. Typically when New England is affected by a hurricane, the area is on the backside of the storm system. Due to the counterclockwise rotation of these storms, this results in a wind blowing from the northeast. Before Franklin's discovery, it was assumed that these storms move northeast to southwest because the wind is from the northeast. Franklin discovers that this often is not true. This re-

1743 *(cont.)*	alization lays the foundation for the discovery that cyclones are rotary in nature. The 1743 hurricane is noteworthy also because it is the first accurately measured by scientific instruments. John Winthrop, a professor of natural philosophy at Harvard College, measures the pressure and tide during the storm passage.
1755	Leonhard Euler derives the equation of fluid flow.
1819	Professor Farrar of Harvard University documents the Boston hurricane of September 23, noting that "it appears to have been a moving vortex and not the rushing forward of the atmosphere." Farrar then describes the extent of the storm damage, the veering of the wind, and how it turned in opposite directions at Boston and New York, as well as the different times of impact on these two cities.
1831	William Redfield makes the first comprehensive analysis of the cyclonic rotation of hurricanes, noting that winds rotate counterclockwise around hurricanes "in the form of a great whirlwind." (In the Southern Hemisphere, the rotation is clockwise; see comments in the year 1835.)
1835	G. G. deCoriolis publishes a paper describing how the earth's rotation causes cyclones (including hurricanes) to rotate counterclockwise in the Northern Hemisphere. This apparent force caused by the earth's rotation, known as the *Coriolis force*, deflects the wind to the right of its intended path in the Northern Hemisphere. In the Southern Hemisphere, the Coriolis force deflects wind to the left of its intended path; therefore, all cyclones in the Southern Hemisphere rotate clockwise.
1839	Henry Piddington coins the word *cyclone*, from the Greek word "kyklos," meaning "coil of the snake," to describe all rotary storms.

1845   The first telegraph line is completed between Washington, D.C., and Boston on April 1, which provides the necessary infrastructure for rapid communication of weather observations. Joseph Henry, secretary of the new Smithsonian Institution, recognizes the opportunity to create an observational weather network based on this technology. In 1848, Henry issues a press release requesting volunteers to transmit weather data to the Smithsonian. By 1849, there are 150 volunteers and by 1860 there are 500 volunteers. This network paves the way for the first government-sponsored synchronous weather observations in 1870.

1847   The first American hurricane warning system is established by Lt. Col. William Reid of the Royal Engineers of England while on duty in Barbados. His warning system is primarily based on pressure measurements.

1870   President Ulysses S. Grant signs a joint resolution of Congress authorizing the secretary of war to organize a national meteorological service under the auspices of the Army Signal Corps on February 9. Later that year, on November 1, the first synchronous weather observations ever taken in the United States are made at 22 stations and telegraphed to Washington. This service also attempts to establish a system for receiving daily observations in the Caribbean to monitor hurricane activity. Originally called "The Division of Telegrams and Reports for the Benefit of Commerce," this agency is the forerunner of the United States Weather Bureau, later called the National Weather Service.

1873   The Army Signal Corps issues its first hurricane warning on August 23 for New England and the mid-Atlantic states.

1874   The Army Signal Corps creates its first weather map depicting a hurricane, for a storm located off the

1874 (cont.)	coast between Savannah, Georgia, and Jacksonville, Florida.
1875	Father Benito Viñes, director of Belen College in Havana, Cuba, develops the first systematic scheme for hurricane forecasts and warnings. This scheme is based on deductions made from observers of sea swells and the motion of high clouds that flow out from the storm center, often indicating the approach of a hurricane. He issues the first forecast in 1875 for a hurricane that hits Hispañola and Cuba, then begins issuing hurricane warnings routinely for this region. The hurricane system consists of a pony express between isolated villages and an organization of hundreds of observers around the Cuban coastline.
1890	The weather service is made a civilian agency on October 1 after two decades of public dissatisfaction with the Army Signal Corps' weather forecasting. For example, the Signal Corps failed to issue a warning for the 1875 hurricane which destroyed Indianola, Texas, and killed 176 people. There was also internal strife, an embezzlement scandal, and a general lack of interest for continuing the labor-intensive observational network because it interfered with military duties. President Benjamin Harrison names the new civilian agency the United States Weather Bureau.
1898	A concerted effort to establish a comprehensive hurricane warning system for North America is begun in preparation for the Spanish-American War at the urging of Willis Moore, the chief of the U.S. Weather Bureau. Fearing for the safety of the American navy, Moore voices his concerns to President McKinley, recalling as an example the 1896 hurricane, which killed 114 people from Florida to Pennsylvania. Impressed by Moore's concerns, President McKinley states that he fears a hurricane more than the Spanish Navy. In response, Congress authorizes funds to establish observation stations in the islands of the Caribbean. A Weather Bureau center is established in Kingston, Jamaica, to issue hurricane forecasts for the West Indies.

After the war is over, additional observing stations are added, and the headquarters are transferred to Havana, Cuba. For political and competitive reasons, all weather statements from Cuban meteorologists are banned. Only weather statements by the Weather Bureau are allowed to be telegraphed—a tragic mistake two years later.

1900   The first serious study of hurricanes by the U.S. Weather Bureau is published, titled "West Indian Hurricanes," by E. B. Garriot. The main contribution is an update of hurricane statistics, although quotes from Redfield, Viñes, and others are included, as well as hurricane tracks for 1875–1900.

A major hurricane strikes Galveston, Texas, killing at least 6,000 people on the island and 2,000 more in nearby coastal regions. Thinking the storm is in the Atlantic, no hurricane warning is issued. Observations from Cuban meteorologists about a hurricane in the Gulf are ignored. This hurricane is still the worst natural disaster in U.S. history in terms of the number of fatalities.

1902   The Cuban hurricane forecasting office is moved to Washington, D.C., in the hope it will improve the U.S. warning system.

Guglielmo Marconi invents wireless telegraphy (radio), which becomes available for ships at sea. On December 3, 1905, the first marine weather report is broadcast at sea, and on August 26, 1909, the SS *Cartago,* near the coast of Yucatan, radios the first ship report about an existing hurricane.

1913   O. L. Fassig publishes a U.S. Weather Bureau technical report called "Hurricanes of the West Indies," which contains statistics and information on hurricane characteristics and swells.

1919   Japanese meteorologist Sakuhei Fujiwhara notes that hurricanes tend to follow the large-scale currents in which they are embedded.

1921   Sakuhei Fujiwhara notes that when two hurricanes get within 850 miles of each other, they begin to move towards each other and rotate cyclonically about a common midpoint between them. This binary interaction between two close hurricanes becomes knows as the "Fujiwhara effect."

1922   Lewis Richardson describes how solutions to meteorological equations may be approximated for weather prediction. His first forecast attempt using this technique fails, but this groundbreaking research lays the foundation for the first successful numerical weather forecasts decades later when computers are invented.

Edward H. Bowie publishes an article on the formation and movement of West Indian hurricanes. He states that hurricanes move in the general motion of the air surrounding it and that they tend to move clockwise around the periphery of a semipermanent feature in the central Atlantic known as the Bermuda high pressure system.

1924   C. L. Mitchell publishes "West Indian Hurricanes and Other Tropical Cyclones of the North Atlantic Ocean" in the American Meteorological Society journal *Monthly Weather Review*, which tracks all tropical storms and hurricanes from 1887 to 1923, with statistics on their frequency, formation points, and motion. This is the most comprehensive study up to that time.

1926   Isaac Cline, the Weather Bureau meteorologist in charge on Galveston Island during the 1900 Galveston Hurricane, publishes *Tropical Cyclones*, which becomes the most authoritative book in the United States on hurricanes at that time. It contains considerable original work on hurricane rainfall, tides, and waves, with the focus on Gulf of Mexico storms.

1935   Congress appropriates $80,000 to revamp the hurricane warning service. New local hurricane forecast centers are established in Jacksonville, Florida; New Orleans; San Juan, Puerto Rico; and later in Boston. A 24-hour hurricane teletype system network is set up from Wilmington, North Carolina, to Brownsville, Texas, so that no more hurricanes are "lost" for several days. These actions are motivated by discontent from coastal communities who feel the Washington, D.C., Weather Bureau hurricane warning service is inadequate. For example, warnings were issued only two hours before the devastating 1926 hurricane hit Miami; even worse, the warning was issued at 11 P.M., after most residents had gone to sleep. Coastal residents also feel Washington, D.C., lacks sensitivity to hurricane problems because the capital city is rarely affected by hurricanes. The best example occurred in August 1934, when the Weather Bureau issued a hurricane warning for the Texas coast one Sunday afternoon. Because no additional observations would be available until 7 P.M. and because the forecaster's shift had ended, the forecaster went home, planning to return to work at 7 P.M. when more observations would be available. During his absence, an anxious Galveston Chamber of Commerce wired the Washington Weather Bureau for an update, and the map plotter on duty honestly but imprudently wired back, "Forecaster on golf course —unable to contact."

Bernhard Haurwitz provides evidence that hurricanes extend from the surface to 20 miles aloft and that the air is anomalously warm at the top of a hurricane.

1937   A nationwide radiosonde network is established. A *radiosonde* is an instrument attached to a balloon that measures the vertical profile of temperature, moisture, pressure, and wind. Forecasters quickly realize that the mid- and upper-level winds steer hurricanes, providing valuable information for track forecasts.

1940   Gordon Dunn publishes a paper showing that most Atlantic tropical storms and hurricanes form from tropical waves—not because of cold fronts, which is

1940
(cont.)
the popular theory at that time. This paper is also the first to document and describe tropical waves.

1942
Weather radar is developed, which detects precipitation by transmitting and receiving reflected radiation (commonly microwave and radio) from raindrops.

1943
Because of a "dare" from a fellow officer, Army Air Corps Col. Joseph P. Duckworth flies the first intentional flight into a hurricane off the coast of Galveston on July 27. Duckworth flies into the eye of the storm twice that day, once with a navigator and again with a weather officer.

Because the busy Jacksonville office lacks sufficient personnel, this hurricane forecast center is moved to Miami to establish a joint hurricane warning service with the Air Corps and the Navy.

1944
Horace Robert Byers shows in his classic book *General Meteorology* that heat and moisture transmitted from the ocean must balance adiabatic (occurring without loss or gain of heat) expansional cooling as the pressure in a hurricane falls. He shows that if air temperature cools too much, the clouds in a hurricane will dissipate and the storm will decay. Thus, hurricanes develop over the ocean, because over land they lack the necessary heat and moisture energy sources. This discovery also explains why hurricanes weaken rapidly upon landfall or when they move over cold water.

World War II radar observations show that rainfall in hurricanes occurs in well-defined spiral bands rather than being uniformly distributed throughout the storm. The first radar picture of a hurricane eye is taken at the U.S. Naval Air Station, Lakehurst, New Jersey, in September.

Herbert Riehl and Major Robert Schafer publish a paper showing that hurricanes do not develop or intensify if the vertical wind shear is too strong.

The Wyman-Woodcock Expedition to the Caribbean Sea occurs, the first of a long series of Woods Hole Oceanographic Institute programs designed to provide insight into the structure of the tropical atmosphere.

1945   Motivated by the need to compute trajectories of bombs and cannon shells during World War II, scientists at the Ballistics Research Laboratory in Maryland develop the first computer, called the Electronic Numerical Integrator and Computer (ENIAC). It is a huge machine measuring more than 100 feet long and 10 feet high and requiring 18,000 vacuum tubes, 70,000 resistors, 10,000 capacitors, and 6,000 switches. This primitive machine makes it possible to perform sophisticated calculations with remarkable speed.

1946   The Navy, with some assistance from the Air Force, begins organized reconnaissance flights scouting for hurricanes. Two squadrons are sent to the West Pacific and one squadron to Miami. These units later become known as the "Hurricane Hunters." The motivation for these flights was the massive damage incurred by the U.S. Navy during World War II from two typhoons in the Pacific. In December 1944 the U.S. Naval fleet, commanded by Admiral William "Bull" Halsey, was going to the Philippine Sea to support the invasion of the Philippine Islands. The meteorologist in charge, Commander George F. Kosco, used what very limited information he had to make a forecast upon which Halsey could steer the fleet. No one knew it at the time, but a scout plane had information about a typhoon that would have altered the fleet's path. The fleet inadvertently sailed into Typhoon Cobra, which had winds up to 130 mph and 70-foot waves. Three destroyers were sunk, 150 carrier-borne aircraft were destroyed, and 790 sailors were killed. After this incident, Halsey requested dedicated reconnaissance flights for hurricanes but was turned down because such heavy-duty aircraft were needed for the war effort. A similar incident

1946
(cont.)
again occurred to Halsey's fleet in June 1945 en route to an invasion of Okinawa, and the Navy then agrees to establish the requested squadron. The reconnaissance team's primary mission is to document the current location, structure, and intensity of hurricanes. Before satellites, reconnaissance flights also search for tropical disturbances.

John von Neumann organizes the Electronic Computer Project at the Institute for Advanced Study (IAS) in Princeton, New Jersey. The goal of this project is to design and build an electronic computer that will exceed the power and capabilities of existing computers, such as ENIAC. In 1948, Jule Charney forms the Meteorology Group at IAS, whose goal is to use this computer for weather forecasting.

1948
Erik Palmen demonstrates that hurricanes do not form over water colder than 80°F, which shows that warm water is a requirement for hurricane formation and intensification.

1950
IAS researchers accurately predict the development of a Thanksgiving snowstorm on a computer. More impressively, the computer takes only 48 minutes to complete the forecast.

1954
In October, pictures taken from the high altitude rocket Aerobee launched by the Navy Research Laboratory reveal a well-developed hurricane off the Texas coast. The discovery of this storm, which had not been detected by the conventional weather-observing network, convinced the U.S. government that routine meteorological observations from higher altitudes were needed. This resulted in the Vanguard satellite program, which began in 1955. Satellites represent the single greatest advancement in observing weather over the tropics.

Inspired by the success of the 1950 IAS computer forecast, the National Weather Service, Air Force, and Navy join forces to form the Joint Numerical Weather

Prediction Group (JNWPU). This group pioneers operational weather forecasts on computers, which become routine in 1955 and gradually improve each year thereafter. In 1958, JNWPU divides into three organizations, which still exist today: the National Meteorological Center, created for civilian needs and now called National Centers for Environmental Prediction (NCEP); the Global Weather Central, created for Air Force needs; and the Fleet Numerical Oceanography Center, created for Navy needs and now called the Fleet Numerical Meteorology and Oceanography Center. All three units, particularly NCEP, interact with the National Hurricane Center and operational forecast centers worldwide for day-to-day operational situations. Also, IAS spawns a new unit now called the Geophysical Fluid Dynamics Laboratory (GFDL), which will later develop the most skillful hurricane track model in the Atlantic.

1955   The Miami Weather Bureau office is officially designated the National Hurricane Center. Gordon Dunn is the first official NHC director, although many, including Dunn, recognize as the "honorary" first director the well-respected Miami forecaster Grady Norton, who died in 1954 while performing forecast duties during Hurricane Hazel.

A new radar is installed at Cape Hatteras, North Carolina, and all three storms that make landfall that year pass within its range, demonstrating the usefulness of radar for tracking hurricanes. By the early 1960s, radars thoroughly cover the U.S. coastline from Brownsville, Texas, to Eastport, Maine, known as the WSR-57 network.

1956   The National Hurricane Research Laboratory (NHRL) is established. Six significant hurricanes hit the northeastern United States in 1954 and 1955 (Carol, Edna, and Hazel in 1954; Connie, Diane, and Ione in 1955) and caused massive damage. With their awareness of hurricanes heightened by these storms, members of Congress authorize funding for NHRL.

1956
(cont.)

Its goals are to examine the structure of all stages of hurricanes and to determine important parameters for hurricane forecasting using instrumented planes and other data sources. Originally located in West Palm Beach, Florida, it is moved to Miami in the same building as NHC. By the late 1990s the lab is called the Hurricane Research Division and is located at Virginia Key, Florida, just east of Miami. This lab's research flights have provided our most complete understanding of the three-dimensional structure of hurricanes.

Herbert Riehl and colleagues develop the first track forecasting scheme based on empirical statistics. This technique assembles past meteorological observations and develops an equation that predicts what a current storm may do based on similar past situations. The advantage of empirical equations is that they are easy to develop and simple to run on computers. The disadvantage is that the atmosphere responds in a much more complicated fashion than suggested by simple statistics, and therefore statistical schemes often do not work well, especially in unusual situations or when a hurricane is rapidly changing track or intensity. Nevertheless, many schemes are devised over the years using statistics.

J. Adem publishes a paper demonstrating that a vortex could slowly propagate to the northwest even in the absence of a steering current, which becomes known as the "beta effect." This discovery shows that hurricane motion is more complicated than a vortex simply following a steering current, and that complex interactions between the hurricane and the environment also determine a hurricane's track.

1959

On May 1, weather forecasting elements of the Navy and Air Force in the western North Pacific are combined into a single tropical cyclone warning center, the Joint Typhoon Warning Center (JTWC) in Guam. Over the years, the role of JTWC grows to include the entire Indian and western Pacific Oceans.

1960   The first experimental weather satellite, TIROS 1, is launched by the United States on April 1 and provides cloud cover photography. Its potential for hurricane study and forecasting becomes obvious in April when it photographs a previously unreported hurricane 800 miles east of Brisbane, Australia.

1962   Project STORMFURY, a study to assess whether people can modify hurricanes by introducing artificial ice nuclei into the eyewall region, begins at NHRL.

1963   Edward Lorenz publishes a landmark paper showing there are inherent limits to the predictability of weather. Because weather forecasts are sensitive to the accuracy of weather observations fed into computers, even slight changes in the decimal point of an observation will yield completely different long-term forecasts. Because measurements of temperature, wind, and moisture will always contain some small errors, long-term weather forecasts will contain large errors (although short-term forecasts—less than 2 days ahead—generally will be accurate). This discovery started a new science called "chaos theory," the study of unstable aperiodic behavior in nondeterministic nonlinear dynamical systems. Because the equations governing the atmosphere are nondeterministic, weather forecasts will always contain errors that grow with time, eventually leading to completely inaccurate long-term forecasts. In theory, the predictability limit is 10 days, but in practice it is about 4–5 days because any additional errors introduced into the computer models accelerate forecast inaccuracies. These additional errors are caused by (1) data gaps in certain parts of the world, especially over the oceans; (2) incomplete understanding of certain atmospheric features, such as clouds, radiation, and rain; and (3) inaccurate treatment of the equations themselves. As a result, forecast errors grow faster in time, especially after 2 days. Chaos theory also applies to hurricanes, and some research has shown that 24-hour track forecasts will always be at least 75 miles off from the actual landfall, on average, due to these limits in predictability.

1964   The weather satellite Nimbus 1 is launched on August 24. This is the first satellite to produce high-quality photographs taken at night.

1965   George Cressman, the director of the Weather Bureau, issues a plan to improve the hurricane warning service by concentrating the responsibilities at the National Hurricane Center in Miami. This plan includes increasing public awareness, providing more reliable service, mitigating the cost of hurricane destruction, and minimizing excessive preparation costs that result from overwarning. The plan also calls for "hurricane specialists" who issue official track and intensity forecasts for the Atlantic and Eastern Pacific oceans. This plan essentially describes the mission and forecast staff at NHC today. Cressman's plan also initiates the first of many administrative changes involving forecast offices in New Orleans, San Juan, Boston, San Francisco, Honolulu, and Washington, D.C., during the next 15 years. Details are contained in Sheets (1990).

1966   Major satellite advancements occur this year. ESSA-1 is launched February 3 and becomes the first operational satellite, providing pictures from two wide-angle television cameras. Twenty-five days later ESSA-2 is launched, providing automated photographs using Automatic Picture Transmission (APT) cameras. Known as *polar-orbiting satellites,* these spacecraft follow a north-south orbit around the earth's poles, thus providing pictures centered on different longitudes each hour as the earth rotates underneath them. Polar orbiting satellites are important research tools because they provide high-resolution pictures. They are also used to infer (but not directly measure) cloud top heights, moisture profiles, temperature profiles, sea surface temperatures, and wind.

            Polar-orbiting satellites, however, lack temporal continuity because the spacecraft will pass over a particular location only twice a day. The solution is to launch a satellite into orbit so that its speed is about the same as the rotating earth, enabling the satellite to remain over the same location and thus provide continuous coverage. These are known as *geosynchronous satellites*. On December 6 the first Applications Technology Satellite (ATS) is launched in geosynchronous orbit. The ATS satellite also has a new camera based on "spin-scan" technology developed by satellite pioneer Verner Suomi at the University of Wisconsin. This technology later evolves into the Geostationary Operational Environmental Satellite (GOES) program, which provides nearly continuous pictures of the entire hemisphere from 60 degrees N to 60 degrees S.

1968   Robert Simpson replaces the retiring Gordon Dunn as director of NHC. Simpson places a renewed emphasis upon research and development at NHC, including the development of satellite applications and statistical equations for predicting hurricane tracks.

1969   Vic Ooyama makes the first two-dimensional computer simulation of a symmetrical hurricane. Ooyama also hypothesizes that hurricane genesis occurs as a cooperative interaction between cloud elements and a tropical disturbance—a theory that becomes known as Conditional Instability of the Second Kind.

1970   The first operational hurricane computer model, called the SANBAR model, becomes operational. Over the years other numerical models become available, including the Moveable Fine Mesh model from 1976 to 1987, the Quasi-Lagrangian Model from 1988 to 1993, the Beta and Advection Model (1989 to current), the VICBAR model (1989 to current), and the Geophysical Fluid Dynamics Laboratory Model (1995 to current).

The National Oceanographic and Atmospheric Administration (NOAA) is created, and the Weather Bureau is renamed the National Weather Service.

1971   Richard Anthes makes the first three-dimensional computer simulation of a hurricane.

1972   The National Data Buoy Center is formed. Its mission is to establish and maintain a buoy network for the coastal United States, Atlantic Ocean, and Gulf of Mexico. Buoys provide key data about tropical systems that would otherwise be unavailable in data-sparse regions.

Charles Neumann develops the CLIPER (CLImatology and PERsistence) statistical equation for predicting storm tracks, which is still used today. CLIPER is considered a benchmark for the evaluation of predictions schemes and models; if the scheme has smaller errors than CLIPER, it is considered skillful.

Vern Dvorak develops a methodology for estimating hurricane intensity based on the storm's cloud pattern as inferred by satellite. This methodology, later refined in 1984 to include infrared pictures and known as the *Dvorak technique,* is still used worldwide today when no direct measurements of hurricane intensity are available.

1973   Neil Frank replaces the retiring Robert Simpson as NHC director. Frank places a renewed emphasis on hurricane preparedness and is particularly skillful at using the media to motivate people in hurricane-threatened areas to respond appropriately.

1974   The U.S. Navy stops flights into hurricanes, leaving only the Air Force to perform routine aircraft reconnaissance.

The GARP (Global Atmospheric Research Program) Atlantic Tropical Experiment (GATE) begins. This project is the first full-phase tropical weather field program and involves 66 countries, 39 ships, and 13 aircraft. This experiment, which occurs in the eastern tropical Atlantic Ocean, provides some of the most complete and well-sampled observations of the tropical atmosphere. These observations allow better understanding of cloud structure, thunderstorms, radiative processes, and turbulence in the tropics. It also produces some of the most detailed data of tropical waves propagating off of Africa, generating new knowledge about the genesis of hurricanes.

1975   The first Geostationary Operational Environmental Satellite (GOES) is launched on October 16. Geostationary satellites gather pictures and other data about the atmosphere in the Atlantic and East Pacific Oceans and send this information to ground stations for use by meteorologists. The GOES satellites are in "geostationary" (also called "geosynchronous") orbit, which means they move at the same speed as the earth's rotation and therefore remain in the same location above the earth. This positioning is advantageous because one can obtain continuous pictures and measurements of the earth in that region.

1977   Meteosat, the first European meteorological satellite to be placed in geostationary orbit, is launched by the United States on November 23. Meteosat provides the first continuous pictures of tropical waves as they form over Africa and propagate westward, and it is still the most important tool for monitoring tropical waves.

The first Geostationary Meteorological Satellite (GMS) satellite is launched on July 14, providing satellite information of the west Pacific region.

1978   The experimental Seasat satellite is launched, containing a new instrument called a scatterometer. A scatterometer emits microwave radiation toward the ocean, and the amount of radiation scattered by ocean waves is proportional to wind speed, thereby indirectly measuring wind speed. Complex algorithms also give wind direction. Based on the success of this technique, later satellites are launched that give wind speed, such as the Special Sensor Microwave/Imagers (SSM/I), and wind speed and di-

1978 *(cont.)*	rection, such as the scatterometers called the European Research Satellites (ERS). The Japanese also launch a satellite called ADEOS that contains the NASA SCATterometer (NSCAT) in 1997, but unfortunately the ADEOS satellite fails within two months. The Quick Scatterometer (QuikScat) is launched in 1999 to replace NSCAT. Scatterometers are important tools for hurricane forecasters monitoring tropical disturbances, as well as for measuring winds on the periphery of hurricanes.
1979	Brian Jarvinen and Charles Neumann develop the statistical scheme SHIFOR (Statistical Hurricane Intensity FORecast) for predicting intensity change. Like CLIPER, SHIFOR is used to assess the skill of other intensity prediction schemes.
1982	HRD begins its Synoptic Flow Experiment. In this experiment, HRD research flights drop many dropsondes within 600 miles of a hurricane's center, obtaining critical temperature, moisture, and wind data required for improved track forecasts. HRD samples 18 hurricanes from 1982–1993, and the experiment successfully shows that extensive measurements of the hurricane environment could yield significantly improved track forecasts by computer models. These results lead NOAA to purchase a high-altitude jet capable of taking meteorological measurements over a large area—the Gulfstream IV (see the year 1997).
1983	A workstation connected to the University of Wisconsin Man computer Interactive Data Access System (McIDAS) is installed at NHC. This system processes, displays, enhances, and animates satellite data and can also overlay the imagery with meteorological measurements. McIDAS is still actively used in the late 1990s by many governmental and university institutions, including NHC. Also included with McIDAS are "cloud-drift winds," which are computed by measuring cloud motion over a given distance. This provides NHC with abundant wind in-

formation in normally data-void regions and is particularly helpful for track forecasts.

William Gray discovers that fewer Atlantic tropical storms and hurricanes occur in El Niño years. A year later Gray begins forecasting the number of tropical storms and hurricanes that will occur months in advance. Gray considers El Niño and such other atmospheric phenomena as the Quasi-Biennial Oscillation, wind shear, and Caribbean surface pressure in making his seasonal hurricane prediction. Gray's annual forecasts demonstrate skill and are issued every year thereafter.

Project STORMFURY ends.

1984   Doppler radar is added to the HRD research flights, enabling scientists to measure the three-dimensional wind structure of hurricanes. Doppler radar uses the *Doppler effect* to measure motions of objects toward or away from the instrument. The Doppler effect is a shift in wavelength of radiation emitted or reflected from an object moving toward or away from the observer. Doppler radar translates the motion of air particles, cloud droplets, and raindrops into wind speed and direction.

1987   Due to budget concerns, the Air Force stops reconnaissance flights in the western North Pacific, leaving only the Atlantic Ocean with routine flight monitoring of hurricanes.

The Improved Weather Reconnaissance System (IWRS) becomes operational onboard the Air Force WC-130 aircraft. This system automatically calculates wind, temperature, humidity, pressure, and heights continuously during a flight. When combined with a direct satellite uplink, this system provides a continuous flow of high-density data to the National Hurricane Center. Prior to IWRS and its satellite communication system, critical weather data (e.g., sea level pressure) were calculated manually

1987
(cont.)

using tables or handheld calculators, then transmitted over HF (high frequency) voice radio circuits across thousands of miles to military radio operators, who then passed the data by phone to Miami.

1988

Robert Sheets becomes the NHC director, replacing the retiring Neil Frank. His term is marked by major technological improvements and a continued emphasis on hurricane preparedness issues related to coastal development and growth.

HRD research flight winds are made available operationally to NHC. This provides additional real-time information about a hurricane's structure to NHC.

1990

Mark DeMaria and John Kaplan develop a statistical scheme that incorporates some environmental factors known to affect hurricane intensity, such as wind shear. This scheme is called SHIPS (Statistical Hurricane Intensity Prediction Scheme).

The Tropical Cyclone Motion Experiment (TCM-90) is conducted over the western North Pacific Ocean during August and September 1990. This experiment's objectives are to test several hypotheses regarding hurricane motion, with a focus on the beta effect. Seven typhoons are analyzed every 12 hours throughout the lifetime of each storm. This experiment is followed two years later by the Tropical Cyclone Motion (TCM-92) mini-field experiment in July and August 1992. TCM-92 is to obtain simultaneous aircraft measurements and satellite imagery into typhoons and nearby cloud complexes. This experiment's objectives are to test hypotheses about tropical cyclogenesis in the cloud complexes and how nearby cloud clusters may cause nearby typhoons to deviate from their track. TCM-92 conducts nine flight missions, and the results support the theory that cyclonic rotation in tropical disturbances begins in the mid-levels of the atmosphere.

1991

The Tropical EXperiment in MEXico (TEXMEX) is conducted over the eastern North Pacific Ocean from

July 1 to August 8. The experiment's objectives are to test several hypotheses regarding tropical cyclogenesis. Six cloud clusters with the potential for formation into a tropical storm are investigated, revealing new insight into the rotational development of tropical disturbances.

1993   Radar pictures and dropsonde data from HRD research flights are included in the operational database for NHC and NCEP's use, thus providing more real-time information about a hurricane's cloud pattern and vertical structure.

Installation of the WSR-88 (NEXRAD) Doppler radar network begins to replace the WSR-57 radar network. This network is useful for detecting tornadoes in severe thunderstorms and for measuring the wind field of hurricanes near coastal Doppler radars. The NEXRAD network can also detect precipitation, the rainfall rate, accumulated rainfall, and the location of the rainfall.

1995   High-resolution satellite pictures of hurricanes are taken in one-minute sequences, revealing unsurpassed imagery of hurricane structure.

Bob Sheets retires as NHC's director and is replaced by Dr. Bob Burpee.

The Geophysical Fluid Dynamics Laboratory model becomes operational at NCEP. This model is the most sophisticated as of the late 1990s and generally has the smallest track errors compared to other model guidance.

1997   A spaceborne radar is launched November 27 aboard the Tropical Rainfall Measuring Mission (TRMM) satellite, a joint venture between the United States and Japan. TRMM is dedicated to measuring tropical and subtropical rainfall using the radar and another instrument—a microwave imager—to measure the vertical distribution of precipitation. TRMM has

1997
(cont.)

been used to develop a database of hurricane cloud structure and to assess if rapid vertical growth of eyewall clouds precedes intensification.

Starting with Hurricane Claudette (in 1997) and many subsequent hurricanes in 1998, routine high-altitude reconnaissance observations are taken in the environment surrounding hurricanes by NOAA's new Gulfstream IV jet, providing new information about the steering currents that are influencing hurricane motion. Dozens of dropsondes per flight are deployed in the hope it will improve computer-guided track forecasts.

The Global Positioning System (GPS) dropsonde system is first deployed into a hurricane (Guillermo in the eastern Pacific) during an HRD research flight. This new device measures wind speeds with unprecedented accuracy and has the ability to take observations below 1,500 feet—something the older Omega dropsondes could not do.

1998

The first robotic aircraft (called an *aerosonde*) is successfully flown across the Atlantic Ocean on August 21–22. The aerosonde was developed in Australia under the direction of Dr. Greg Holland of the Bureau of Meteorology. The Atlantic flight is conducted by the University of Washington and the Insitu Group, a U.S. engineering company. The unmanned 29-pound plane departs from St. John's, Newfoundland, Canada, and completes the 2,000-mile trip to the west coast of Benbecula in the Outer Hebrides of Scotland in just over 26 hours. The trip is completed using less than 2 gallons of fuel, continuously measuring meteorological observations the entire time. These drones theoretically can be programmed to fly a fixed path for over 24 hours, continuously taking weather observations that would improve weather forecasts. Their unique aerodynamic structure and extreme light weight also theoretically allow them to withstand hurricane winds and strong updrafts in thunderstorms, thereby providing timely and contin-

uous weather data at less cost than airplane reconnaissance.

NASA and NOAA collaborate in the Third Convection and Moisture Experiment (CAMEX-3) in an effort
to collect comprehensive data on Atlantic hurricanes
using multiple aircraft, including high altitude measurements using the Gulfstream IV and NASA's ER-2
aircraft, with a variety of remote sensing technologies
(including scanning radar altimeter and Doppler
radar). On August 20, Hurricane Bonnie becomes the
first storm to be sampled in this project, using up to
six planes at once (the Gulfstream IV, ER-2, NASA
DC-8, two WP-3D research flights conducted by HRD,
and one WC-130 flown by the Hurricane Hunters).

The largest coordinated measurements to date of
landfalling hurricanes are performed for Hurricanes
Bonnie, Charlie, Earl, and Georges. These include
portable instrumented towers, portable wind profiler, and a portable Doppler radar. Texas Tech University, Clemson University, the University of Okla-
homa, and the University of Alabama at Huntsville
are the principal investigators.

NHC's director Bob Burpee retires and is replaced by
Jerry Jarrell.

Chronology of Some Hurricanes
That Affected History

Fourteenth A fleet from Mongolia, possibly the largest at that
century time, is about to invade (and certainly conquer)
Japan when it is destroyed by a typhoon. This great
fortune for Japan gives rise to the term *kamikaze,* or
divine wind.

1565 A hurricane scatters the French fleet off the Atlantic
coast of North America and allows Spain to capture

1565 (cont.)	France's Fort Caroline, at the present site of Jacksonville, Florida. The French therefore lose their bid for control of North America to the Spanish.
1588	Spain, at war with England and becoming more worried about English sea power, dispatches the largest naval force in history, dubbed "The Invincible Fleet," to invade England. However, two weakening but still potent hurricanes (at the end of their life cycle) hit the Spanish Fleet that summer, destroying half the fleet. Spain loses control of the sea, and therefore its domination of the New World, to England.
1609	The British ship *Sea Venture,* en route to Jamestown, Virginia, is grounded on Bermuda by a hurricane. The stranded people are the island's first inhabitants, who think the island is "paradise," and claim the island for England. This incident inspired Shakespeare to write one of his best plays, *The Tempest.* Today Bermuda is a territory of the United Kingdom.
1640	A hurricane partially destroys a large Dutch fleet poised to attack Havana, Cuba. This disaster helps the Spaniards secure control of Cuba.
1666	England loses a fleet of 17 ships and 2,000 troops in a hurricane near Guadeloupe. The French capture the remaining survivors, resulting in France's control of Guadeloupe through the twentieth century.
1780	The Great Hurricane strikes the Lesser Antilles in the Caribbean Sea on October 10 and is the most deadly hurricane in Western Hemisphere history. Estimates indicate approximately 22,000 deaths, with 9,000 lives lost in Martinique, 4,000–5,000 in St. Eustatius, and 4,326 in Barbados. Thousands of deaths also occur offshore, including great losses in the fleets of Britain and France, who were planning assaults on each other to claim the Antilles Islands. After the Great Hurricane of 1780, the governor of Martinique freed the English soldiers who had become his prisoners, even though the French and English were at

war. He declared that in such disasters all men should feel as brothers.

1837   A large major hurricane, dubbed Racer's Storm, travels along most of the coastline of the Gulf of Mexico and literally fills the Gulf of Mexico with an estimated diameter of 1,600 miles. The storm crosses the Yucatan peninsula, heads north along the Mexico and Texas coast, then turns east over the Louisiana coast before finally making landfall along the Alabama-Florida border. It then exits into the Atlantic just below Charleston, South Carolina. It causes destruction all over the Gulf Coast, wipes out the Mexican Navy, and destroys several U.S. ships. Another ship, the *Racer*, survives the storm, goes to Havana, Cuba, for repairs, and provides valuable information about the storm to William Reid. As a result, this storm is known as Racer's Storm. Off North Carolina, the paddle-wheel steamer *Home*, then hailed as the nation's fastest ship and on pace to set a new speed record during the trip, encounters the hurricane and sinks off the North Carolina coast. There are only two life preservers on the ship. Forty passengers struggle to shore, and the rest (mostly women and children) drown. This event causes Congress to rewrite maritime laws requiring a life preserver for every person on a ship.

1889   In March 1889 war is prevented between the United States and Germany by a typhoon. Late in 1888 a German naval force carries the native chief of the Samoans away and sets up another king in his place. The natives rebel and 22 German soldiers are killed. The Germans retaliate by sending three warships into the island's harbor and shelling a native village, incidentally destroying property of American citizens located there. The Germans also rip down and burn an American flag. The United States sends three warships of its own to protect its citizens' lives and property. On March 16, the U.S. and German fleets are in the harbor, poised to attack should either side provoke the other. However, before any overt actions occur, a typhoon strikes the island.

1889
(cont.)

The six warships and six other merchant ships are pummeled against reefs, sunk, or beached. About 150 people on the ships lose their lives. But during the storm, all three factions—the Germans, Americans, and Samoans—unite as allies and help each other, performing many heroic acts. Afterward, Germany and the United States decide to resolve their differences peacefully. In the resulting Treaty of Berlin of 1889 it is agreed that Samoa will remain an independent territory with its own rulers. This incident also makes the United States realize it is a "major power" in the world, which requires a larger navy to curb other aggressive countries. In other words, the Samoa incident is indirectly responsible for the founding of the modern U.S. Navy.

1898

In 1898 Cuba is part of the oppressive Spanish empire. After the USS *Maine* is sunk under mysterious circumstances in Havana, Cuba, the Americans declare war on Spain. An American regiment called the "Rough Riders" under Lt. Col. Theodore Roosevelt quickly overwhelm the Spanish garrison and liberate Cuba. Lookout posts are established along the Cuban coastline to watch for a Spanish counterattack, but by August 12 the war is over and Cuba is free. The U.S. weather service sees great potential for the Cuban lookout posts and takes them over as part of a hurricane early warning system. They are dismayed to find the outdoor privy (or, in other words, an outhouse) has been stolen from one of the stations. This causes such an uproar that the local, who had stolen it for his own personal use, returns it. Very shortly afterward, on October 2, both the observation post and its privy are destroyed by a hurricane, which is immortalized as the "Privy Hurricane of 1898."

References

Allaby, M., 1998. *A Chronology of Weather.* New York: Facts on File, 154 pp.

Byers, H. R., 1974. *General Meteorology.* New York: McGraw-Hill, 461 pp.

Cline, I. M., 1926. *Tropical Cyclones.* New York: MacMillan Company, 301 pp.

Danielson, E. W., J. Levin, and E. Abrams, 1998. *Meteorology.* WCB/Mc-Graw-Hill Co., 462 pp.

Douglas, M. S., 1976. *Hurricane.* Atlanta, GA: Mockingbird Books, 119 pp.

Dunn, G. E., and B. I. Miller, 1964. *Atlantic Hurricanes.* Baton Rouge: Louisiana State University Press, 377 pp.

Emme, E. M., 1961. *Aeronautics and Astronautics: An American Chronology of Science and Technology in the Exploration of Space, 1915–1960.* Washington, DC: National Aeronautics and Space Administration, 88 pp.

Fassig, O. L., 1913. Hurricanes of the West Indies. Bulletin No. 13. Washington, DC: U.S. Weather Bureau, 28 pp.

Fincher, L., and B. Read, 1999. *The 1943 "Surprise" Hurricane.* Available at *http://www.ci.houston.tx.us/departme/finance/oem/* and the City of Houston Office of Emergency Management.

Fishman, J., and R. Kalish, 1994. *The Weather Revolution—Innovations and Imminent Breakthroughs in Accurate Forecasting.* Plenum Press, 276 pp.

Fleming, J. R., ed., 1996. *Historical Essays on Meteorology.* Boston: The American Meteorological Society, 617 pp. (Chapter 9, written by M. De-Maria, discusses the history of hurricane forecasting for the Atlantic Basin from 1920–1995).

Frisinger, H., 1983. *The History of Meteorology to 1800.* Boston: The American Meteorological Society, 148 pp.

Grice, G. K., 1999. *The Beginning of the National Weather Service: The Signal Years (1870–1891) as Viewed by Early Weather Pioneers.* Available from *http://tgsv5.nws.noaa.gov/pa/special/history* and from the National Weather Service Office of Public Affairs.

Griffith, J. F., 1977. A Chronology of Items of Meteorological Interest. *Bull. Am. Meteor. Soc.*, 58:1058.

Heidorn, K. C. 1978. A Chronology of Important Events in the History of Air Pollution Meteorology to 1970. *Bull. Am. Meteor. Soc.*, 59:1589.

Hughes, P., 1987. Hurricanes Haunt Our History. *Weatherwise*, 40:134–140.

———, 1990. The Great Galveston Hurricane. *Weatherwise*, 43:190–198.

Jennings, G., 1970. *The Killer Storms: Hurricanes, Typhoons, and Tornadoes.* New York: J. B. Lippincott Co., 207 pp.

Larson, E., 1999. *Isaac's Storm.* New York: Crown Publishers, 323 pp.

Ludlum, D. M., 1989. *Early American Hurricanes: 1492–1870.* Boston: American Meteorological Society, 198 pp.

Middleton, W. E. K., 1969. *Invention of the Meteorological Instruments.* Johns Hopkins Press, 362 pp.

Mitchell, C. L., 1924. West Indian Hurricanes and Other Tropical Cyclones of the North Atlantic Ocean. *Mon. Wea. Rev.,* Supplement No. 24, 47 pp.

Neumann, C. J., B. R. Jarvinen, C. J. McAdie, and J. D. Elms, 1993. *Tropical Cyclones of the North Atlantic Ocean, 1871–1992.* Asheville, NC: National Environmental Satellite, Data, and Information Service, National Climatic Data Center, 193 pp.

Sheets, R. C., 1990. The National Hurricane Center—Past, Present, and Future. *Wea. Forecasting,* 5:185–232.

Some Devastating North Atlantic Hurricanes of the Twentieth Century, 1982. Washington, DC: U.S. Dept. of Commerce, National Oceanic and Atmospheric Administration, 16 pp.

U.S. Department of Commerce, 1969. Hurricane Camille, August 14–22, 1969 (Preliminary Report). Environmental Science Services Administration, Weather Bureau, Washington, DC, 58 pp.

Weems, J. E., 1997. *A Weekend in September.* College Station: Texas A & M University Press, 192 pp.

Biographical Sketches 3

Many individuals have contributed to hurricane research, forecasting, and preparedness in the past two centuries. Biographical sketches of past and current hurricane scientists who have made important contributions to the field are included in this chapter. This list should be considered just a sample, because many people have provided key contributions to tropical meteorology, weather forecasting, disaster mitigation, and hurricanes in the last 200 years. Noteworthy books published by the individuals are listed where appropriate; these books are available from several libraries, but a good starting point is the NOAA library. These books may be obtained via interlibrary loan from the NOAA library and are listed on their web site (see http//www.lib.noaa.gov for details).

Richard Anthes (1944–)

Richard A. Anthes, president of the University Corporation for Atmospheric Research (UCAR) since September 1988, is a highly regarded atmospheric scientist, author, educator, and administrator who has contributed considerable research to weather modeling, thunderstorms, and hurricanes. Anthes was born March 9, 1944, in St. Louis, Missouri, and grew up in Waynesboro, Virginia. He knew as a very young child that he wanted to be a meteorologist.

While attending the University of Wisconsin-Madison to earn his bachelor's degree, Anthes pursued his interest in meteorology by working as a student trainee for the U.S. Weather Bureau at the National Oceanic and Atmospheric Administration during the summers of 1962 through 1967. He discovered that an area of particular interest to him during this period was hurricanes. His master's and doctorate theses, completed in 1967 and 1970 respectively from the University of Wisconsin-Madison, reflected this interest. In particular, Anthes's doctorate work resulted in the first three-dimensional computer simulation of a hurricane. From 1968 to 1971, he was also a research meteorologist at NOAA's National Hurricane Research Laboratory.

In 1971, Anthes started teaching and conducting research at Pennsylvania State University, where he attained a full professorship in 1978. During this period, he also acted as a visiting research professor at the Naval Postgraduate School in Monterey, California. Anthes's research group studied a variety of weather issues related to hurricanes and thunderstorms. He frequently published papers on hurricanes, including one book titled *Tropical Cyclones—Their Evolution, Structure, and Effects*.

Because hurricanes and thunderstorms contain small-scale features requiring high-resolution and different equations, Anthes modified the weather model he used for his doctorate to incorporate the proper physics for studying them. Research and development of this model has continued ever since, and several universities and government laboratories are continually improving the model. As of the late 1990s this model, known as the Fifth-Generation NCAR/Penn State Mesoscale Model (MM5), is a state-of-the-art public domain model available from University Corporation for Atmospheric Research. MM5 has been used by hundreds of graduate students and scientists for conducting modeling research and weather forecasts.

The National Center for Atmospheric Research (NCAR) welcomed Anthes in 1981 when he became the director of NCAR's Atmospheric Analysis and Prediction Division. In 1986, Anthes was selected to become the director of NCAR, and in 1988 he was selected to become the president of the University Corporation for Atmospheric Research (UCAR). UCAR is a nonprofit consortium of 61 member universities that awards Ph.D.s in atmospheric and related sciences. UCAR manages NCAR, in addi-

tion to collaborating with many international meteorological institutions through a variety of programs.

Anthes was elected as an American Meteorological Society
(AMS) Fellow in 1979. In 1980, he was the winner of the AMS's
Clarence L. Meisinger Award as a young, promising atmospheric
scientist who had shown outstanding ability in research and
modeling of tropical cyclones and mesoscale meteorology. In
1987, he received the AMSs Jule G. Charney Award for his sustained contributions in theoretical and modeling studies related
to tropical and mesoscale meteorology.

He has participated in or chaired over 30 different national
committees (for such agencies as NASA, NOAA, AMS, NSF, the
National Research Council, and the National Academy of Sciences), including his present chairmanship of the National
Weather Service Modernization Committee. Anthes has published over 90 peer-reviewed articles and books. One book in particular, *Meteorology* (7th ed., 1996) is widely used at colleges and
universities as a general introductory book to the field of meteorology for nonmeteorological majors.

Isaac Monroe Cline (1861–1955)

Isaac Monroe Cline is known for being the Weather Bureau's meteorologist in charge at Galveston Island during the Galveston
hurricane of 1900, which killed 6,000 people. He was also a distinguished forecaster and prolific author.

Cline was born in 1861 and entered the Army Signal Corps
in 1882. At that time, he entered the weather service school at Fort
Myer, Virginia. Some of his instructors were distinguished physi-
cists, such as William Ferrel, T. C. Mendenhal, and Cleveland
Abbe. Upon completing school, Cline was assigned to offices in
Little Rock, Arkansas; Fort Concho, Texas (near San Angelo); and
Abilene, Texas, before being assigned to Galveston, Texas, as the
chief meterologist.

In July 1891, Cline wrote an article for the *Galveston News*
stating that Texas was "exempt" (p. 80) from intense hurricane
landfalls, and that any landfalls would be from weak hurricanes. He reasoned that, since no intense hurricanes had made
landfall on Galveston in recent history, strong hurricanes must
recurve before reaching Galveston. In fact, two deadly hurricanes had hit Indianola (120 miles to the south) in 1875 and
again in 1886, but he unknowingly or deliberately labeled these
as weak storms, and did not mention the fatalities. He also

wrote, "it would be impossible for any cyclone to create a storm wave which could materially injure the city" (p. 84). Cline theorized that the shallow slope of the seabed off Galveston would wear down incoming seas before they struck the city (in fact, the opposite is true) and that the storm surge would spread first over the lowlands surrounding Galveston. It is unclear how this article impacted his judgment of the Galveston hurricane nine years later, but it may have contributed to some facets of the disaster.

On September 5, 1900, a tropical system passed over Cuba and developed into a hurricane in the eastern Gulf of Mexico. The Weather Bureau in Washington, D.C., had mistakenly concluded this system was east of the Florida Keys and would be moving into the Atlantic. Cuban meteorologists had telegraphed that a hurricane had formed in the Gulf, but for political and competitive reasons the Weather Bureau banned all Cuban weather telegraphs. After the disaster, Weather Bureau Chief William Moore released a statement falsely claiming his agency had warned Galveston of the coming hurricane, and many journalists accepted Moore's statement. In fact, Washington never issued a hurricane warning to Galveston.

On September 8, Cline observed increasing ocean swells, stronger winds, and minor flooding, but, based on Weather Bureau statements, Cline concluded it was just an "offspur" of a Florida storm. (Legend states that Cline hitched up his horse and cart and traveled up and down the beach, telling homeowners to move to higher ground and vacationers to go home; this is, however, a myth). The water level had risen several feet and winds had increased rapidly. By 2:30 p.m., a concerned Cline had his brother Joseph (who also worked for the Weather Bureau) telegraph weather observations to Washington D.C. with the statement, "Gulf rising rapidly; half the city now under water." Both brothers then waded home to Isaac's house. It was the last message out of Galveston.

By late afternoon, the full fury of the hurricane hit. That evening Cline's house, already pummeled by the still increasing storm surge, started to float. When his house rolled, he briefly lost consciousness but recovered and found all his family except his pregnant wife, who had been ill in bed. Cline drifted for several hours on one sinking piece of debris after another. With him were his three daughters, his brother Joseph, and a little girl Joseph had pulled from the water. Finally, they came to rest on wreckage which turned out to be right back in their own neighborhood. One month later the body of Cline's wife was found under that very wreckage.

Despite his personal grief, Cline later wrote a technical paper on the Galveston Hurricane titled "Special Report on the Galveston Hurricane of September 8, 1900." Cline was the author of a multitude of meteorological papers and books. The most noteworthy was titled *Tropical Cyclones*, published in 1926, and was the most authoritative book in the U.S. on hurricanes at that time. It contains considerable original work on hurricane rainfall, tides, and waves, with the focus on storms in the Gulf of Mexico. Another noteworthy book is *Storms, Floods, and Sunshine*, which contains two parts. The first part is his memoirs, and the second part describes the characteristics of hurricanes.

Cline also distinguished himself in 1927 during the great Mississippi Valley flood, and for this he later received a special commendation from President Herbert Hoover. Cline spent his later years in New Orleans and died in 1955 at the age of 94.

Mark DeMaria (1955–)

Mark DeMaria was born September 30, 1955, in Middleburg, Pennsylvania, although he spent most of his early life in Miami, Florida. He developed an interest in hurricanes at a young age after witnessing Hurricanes Cleo, Betsy, and Inez in 1964–1966. He obtained a bachelor of science in meteorology from Florida State University in 1977 and a master of science and Ph.D. in atmospheric science in 1979 and 1983, respectively, from Colorado State University. He developed a spectral tropical cyclone model under the direction of Dr. Wayne Schubert for his Ph.D. research. After short periods of employment at the National Center for Atmospheric Research (NCAR) and North Carolina State University, DeMaria returned to Miami in 1987 to work at the Hurricane Research Division (HRD). In 1995 he left the research community to become the chief of the technical support branch at the Tropical Prediction Center/National Hurricane Center in Miami.

The primary emphasis of DeMaria's work is on the development of techniques for hurricane track and intensity forecasting. His early research focused on track forecasting, and he won the Banner Miller Award from the American Meteorological Society in 1985 and again in 1987 for studies on barotropic vortex motion. As part of his work at HRD, DeMaria implemented an experimental tropical cyclone track forecast model (Victor's barotropic model, or VICBAR) using a nested spectral method developed by Dr. K. V. "Vic" Ooyama. The VICBAR model was also used in the first study to demonstrate statistically significant track forecast improvements for cases that included experimental dropsonde data collected in the vicinity of tropical cyclones with the NOAA WP-3D aircraft. In collaboration with John Kaplan of HRD, DeMaria developed a Statistical Hurricane Intensity Prediction Scheme (SHIPS) for the Atlantic basin. In 1996, SHIPS became part of the routine forecast guidance run in real time at the National Hurricane Center. After the 1996 hurricane season, SHIPS was reformulated to more accurately evaluate the forecast predictors. Since that time, SHIPS is the only model that has demonstrated statistically significant intensity forecast skill for Atlantic storms. Also with John Kaplan, DeMaria developed an empirical model for predicting the inland decay of tropical cyclone winds. In cooperation with the Federal Emergency Management Agency (FEMA), results from this algorithm were distributed to many emergency management agencies in the southeast United States for planning purposes. In 1997, he (with John Kaplan) received an NOAA Bronze Medal for this effort.

In late 1998, DeMaria returned to Colorado to become the leader of the NESDIS Regional and Mesoscale Meteorology (RAMM) Team at Colorado State University. He plans to continue his research on tropical cyclone forecasting, with emphasis on the use of remote sensing data. DeMaria has published more than 35 peer-reviewed articles on hurricanes and tropical meteorology, and in 1995 was invited to contribute a review article on the history of hurricane forecasting to a special Diamond Anniversary volume of the American Meteorological Society.

Gordon Dunn (1905–1994)

Gordon Dunn was a weather enthusiast with an uncanny ability to synthesize observations and forecast weather and hurricanes accurately. Dunn was born on August 9, 1905, in Brownsville, Vermont. He developed a fascination with the weather as a young boy, and as a teenager he would hitch rides to the Weather Bureau office 18 miles away to study weather charts and discuss weather events with the station's sole employee. Dunn applied this knowledge to practical weather problems on his father's dairy farm.

Dunn accepted a job as a messenger with the Weather Bureau in Providence, Rhode Island, in 1924, where he took surface observations, carried messages to the Western Union office, and drew the weather map for the morning paper. He transferred to the Weather

Bureau in Tampa, Florida, with a promotion to junior observer two months later. In 1926 he requested relocation to an area where he could complete his college education and transferred to the Weather Bureau's Central Office in Washington, D.C. In this location, he attended George Washington University to pursue an A.B. degree in political science, which he completed in 1932. However, he also took courses in meteorology at other universities whenever possible, such as Florida State University, Massachusetts Institute of Technology, and the University of Chicago. In 1931, he was promoted to a meteorologist in the Washington Weather Bureau.

In 1935 Congress appropriated $80,000 to revamp the Weather Bureau's hurricane warning service. Four new hurricane forecast centers were established, including one in Jacksonville, Florida. Grady Norton was transferred to Jacksonville as the senior forecaster, and Dunn was transferred there as the junior forecaster. Dunn and Norton faced the challenge of forecasting several devastating and deadly hurricanes, including the Labor Day Hurricane (1935) and the New England Hurricane (1938). During this period, Dunn noticed areas of rising and falling pressure that progressed westward every three to four days in the Atlantic Ocean with an inverted, wavelike pressure signal; he also noticed some of these systems became hurricanes. In 1940 Dunn published a paper showing that most Atlantic tropical storms and hurricanes form from these disturbances, called tropical waves (also called easterly waves)—not from cold fronts, which was the popular theory at that time. This paper was also the first to document and describe tropical waves.

In 1939, Dunn was transferred to Chicago, where he soon became the meteorologist-in-charge, a position he held until 1955. Dunn developed close ties between the Weather Bureau and the well-respected University of Chicago meteorology program and is credited with improving working relationships between forecasters and professors. During World War II, Dunn was assigned to Calcutta, India, where he assisted the military in weather forecasts in the China-Burma-India region. The military awarded Dunn the Medal of Freedom for his outstanding weather analyses and forecasts.

In 1955 the Miami Weather Bureau office was officially designated the National Hurricane Center. Gordon Dunn was named the first official NHC director. He continued to work there until his retirement in 1968. During this period, Dunn originated procedures for improving public awareness of the hurricane threat and improved international cooperation, including training and upgrading of meteorological services throughout the Caribbean.

Dunn received many awards for his efforts as the NHC director. The Department of Commerce awarded him its Gold Medal in 1959. In 1966, the University of Miami awarded Dunn an honorary doctorate of science. The American Meteorological Society named him an honorary member in 1992.

Dunn died in south Miami on September 12, 1994.

Russell Elsberry (1941–)

Russell Elsberry grew up in Colorado and attended Colorado State University, where he earned a bachelor of science degree in mechanical engineering in 1963. During his undergraduate studies, he worked as a computational assistant for Herbert Riehl performing tasks related to meteorology. He decided that studying hurricanes was more exciting than mechanical engineering and pursued a Ph.D. in meteorology with Riehl as an advisor. Upon completing his doctorate in 1968, Elsberry became an assistant professor in the Department of Meteorology at the Naval Postgraduate School, where he has remained. Elsberry was promoted to associate professor in 1972, was named professor in 1979, and was awarded distinguished professor status in 1994.

Elsberry has remained very active in hurricane research using observational and numerical modeling approaches, particularly with regard to understanding tropical cyclone motion. His research group has studied how the tropical cyclone circulation interaction with the environment may affect the motion. They have also developed a systematic approach for forecasting tropical cyclone motion that is being extended from the western North Pacific to other tropical cyclone basins. He was technical director of the Office of Naval Research Tropical Cyclone Motion Project, which coordinated a field program called the Tropical Cyclone Motion (TCM-90) Experiment over the western North Pacific in August and September 1990. TCM-90's objectives were to test several hypotheses regarding hurricane motion, with a focus on the environmental interactions. This program was followed two and three years later by the Tropical Cyclone Motion (TCM-92 and TCM-93) mini–field experiments in July and August 1992 and 1993. TCM-92 and TCM-93 were to obtain simultaneous aircraft measurements and satellite imagery into typhoons and nearby cloud complexes. Their objectives were to test hypotheses about how these nearby cloud clusters might cause typhoons to deviate from their track and also how tropical cyclogenesis might occur from the cloud complexes.

Elsberry was editor of two books that provided international overviews on tropical cyclones: *A Global View of Tropical Cyclones* (published in 1987 by the University of Chicago Press) and *Global Perspectives on Tropical Cyclones* (published in 1995 by the World Meteorological Organization). He is serving as the Hurricane Landfall Deputy Lead Scientist for the U.S. Weather Research program. Elsberry has published numerous papers on tropical cyclones and is a fellow of the American Meteorological Society.

Neil Frank (1931–)

Neil Frank is credited with improving public awareness of hurricane preparedness. He attended Southwestern College in Kansas where he played on the basketball team for four years. Encouraged by his coach to pursue a degree in the sciences, Frank earned a bachelor of science degree in chemistry with the intention of being a science and physical education teacher. After college, Frank joined the Air Force and served in Okinawa, Japan, where he became fascinated by the typhoons that frequently affected that region. As a result, he pursued graduate studies in tropical meteorology at Florida State University, where he earned a master of science and Ph.D. in meteorology. In 1961, Frank began work at the National Hurricane Center as a forecaster and in 1974 became its director. In 1987, Frank left government employment for his current position as chief meteorologist at KHOU-TV in Houston, Texas.

Frank's interest in public awareness began in 1965 while investigating Hurricane Betsy's storm surge in Miami. While surveying a damaged area, Frank was approached by a muddy survivor of the storm surge who angrily asked why no one at NHC was informing people about hurricanes. Based on this experience, Frank emphasized hurricane preparedness during his tenure as NHC director, frequently giving slide show presentations to coastal audiences, presenting background hurricane briefings to the press, and effectively using the news media as a means of disseminating information when a hurricane threatened the United States. Frank also encouraged upgrades to local community hurricane preparedness plans and employed storm surge models to graphically demonstrate possible hurricane landfall scenarios. He served as chairman of an International

Hurricane Committee, which coordinated hurricane warning procedures for North American countries. In 1987 he testified to the Senate Commerce, Science, and Transportation Committee on hurricane preparedness.

William M. Gray (1929–)

Probably the best-known hurricane scientist among the general public today, William M. Gray is frequently quoted by media sources throughout the country. He was born October, 9, 1929, in Detroit, Michigan, and attended high school in Washington, D.C. Hours after graduating from high school, Gray left for Florida to join the Washington Senators training camp, but he did not make the team. Instead, he pitched baseball in college while pursuing a degree in geography at Wilson Teachers College (1948–1950) and George Washington University (1950–1953). During this time, Gray suffered a leg injury which forced him to abandon the pursuit of a baseball career. After earning his degree, he joined the Air Force, where he used his scientific background to forecast the weather from 1954 to 1957 in the Azores Islands and England. Gray then attended the University of Chicago to study meteorology under the mentorship of Dr. Herbert Riehl, where he earned a master of science in 1959 and a Ph.D. in 1964. However, much of the Ph.D. work was actually completed as a faculty member at Colorado State University, where Riehl started a new meteorology program in 1961 and invited Gray along.

From 1961 to 1983, Gray developed a reputation as one of the premier hurricane researchers in the world. He formulated the general conditions necessary for tropical genesis and developed detailed data sets of hurricane structure by compositing observations from data-sparse regions and reconnaissance flights. In 1983 Gray discovered that fewer Atlantic hurricanes occur during El Niño years. Although he was well respected among his peers, he was relatively unknown to the general public during this period.

In 1984, Gray announced a scheme for forecasting the number of tropical storms and hurricanes in the Atlantic. Gray also boldly made this forecast available to the general public. The forecasts initially generated skepticism from scientists and nonscientists alike. However, with each year the forecasts showed skill, making Gray a household name among coastal residents. Gray's success is considered one of the most influential works in the field of *seasonal predictions*, in which a forecaster predicts below average, average, or above average weather conditions months in advance; the weather conditions may be temperature, rainfall, hurricane activity, and so on. Gray's seasonal hurricane forecasts are based on the relationship between hurricane activity and such factors as El Niño, African rainfall, sea surface temperature, and Caribbean surface pressure (see Chapter 1). These forecasts are initially issued in December and are updated in April, June, and August of each year. These predictions are widely distributed by the press, and on the website *http://tropical.atmos.colostate.edu.*

Gray has served on or chaired more than 20 important committees and panels and is author or coauthor of 83 reviewed publications. He has served as a visiting scientist in many countries, including Japan, China, and England. Among the awards he has received are the Jule G. Charney Award (1993) and the Neil Frank Award of the National Hurricane Conference (1995). He was the ABC Television "Person of the Week" in September 1995 after accurately predicting a very active Atlantic hurricane season. He was an invited lecturer at the Twelfth WMO Congress, Geneva, in June, 1995, an honorary award given to senior scientists in recognition of lifetime research achievements. Gray has graduated 37 students with a master of science and 17 with a Ph.D., many of whom have also enjoyed successful meteorology careers. Gray is a fellow of the American Meteorological Society.

Greg Holland (1948–)

Greg Holland is an Australian meteorologist who has studied many aspects of tropical meteorology and tropical cyclones with a unique Southern Hemisphere perspective. Born August 15, 1948, Holland obtained a bachelor of science in physics with first class honors in mathematics and a minor in mathematics from the University of New South Wales in 1972. He joined the Australian Bureau of Meteorology as a cadet while still at the university and also received a Diploma of Meteorology from the Central Training School in the Bureau of Meteorology in 1972. From 1973 to 1975, Holland was a forecaster for the Darwin Tropical Analysis and Regional Forecasting Centre, then a lecturer for the Central Training School in the Bureau of Meteorology from 1975–1977 in Melbourne. From 1978 to 1984, the Bureau of Meteorology sent Holland to Colorado State University, where he obtained a master of science in 1981 and a Ph.D. in 1983 with William Gray as his advisor. Holland has remained at the Bureau of Meteorology under a variety of titles. He is also an Honorary Associate of the

Faculty at Monash University and an Honorary Professor at the University of New South Wales, under which he advises several Ph.D. students.

Holland has participated in numerous tropical cyclone research activities during his career. He has studied mechanisms for their formation, intensification, and motion. He has participated in all major field experiments and panel discussions in the last 20 years, publishing numerous papers and technical reports along the way. He is technical director of the Aerosonde Project, which is developing robotic aircraft (called aerosondes) that continuously measure temperature, wind, and moisture in data-sparse oceanic regions. One goal is to fly these aerosondes into hurricanes to obtain key information for research and forecasting. Holland is a fellow of the American Meteorological Society.

Yoshio Kurihara (1930–)

Yoshio Kurihara has been a leader in theoretical and operational hurricane modeling for the past 25 years. Kurihara was born on October 24, 1930, in Japan. He obtained a bachelor of arts degree at the University of Tokyo in 1953, and a Ph.D. at the University of Tokyo in 1962. During this period, he worked at the Japan Meteorological Agency in Tokyo from 1953 to 1959 and at the Meteorological Research Institute in Tokyo from 1959 to 1963 and from 1965 to 1967, interrupted by temporary employment at the Geophysical Fluid Dynamics Laboratory (GFDL) from 1963 to 1965. In 1967, Kurihara returned to GFDL, where he remained until his retirement in 1998. Kurihara then returned to Japan, where he is conducting tropical cyclone research at the Frontier Research System for Global Change in Tokyo.

Kurihara was the group leader of the hurricane research project at GFDL that was initiated in 1970. Kurihara's group developed one of the first three-dimensional models of hurricanes and has conducted a number of numerical simulation studies in order to understand the basic mechanism of hurricane evolution. Many of these early studies dealt with spiral band structure and hurricane landfall. Although this model was originally developed for pure research, tests from 1986 to 1992 showed that their model was adept at operational track forecasts. The GFDL was tested semi-operationally in 1993. Unlike several other models, the GFDL model correctly predicted that Hurricane Emily would take a sharp turn from North Carolina's Outer Banks and just graze the area. The GFDL model has officially been used as a hurricane forecast model since 1995 and generally has the smallest track errors compared to other model guidance. A similar model is used by the Navy in the western Pacific and is called the GFDN model.

Kurihara has received the following awards during his career: the Meteorological Society of Japan Award in 1975, the Distinguished Authorship Award from NOAA in 1983, the Banner Miller Award from the American Meteorological Society in 1984 and in 1997, the Environmental Research Laboratories Distinguished Authorship Award from NOAA in 1992, the Department of Commerce Gold Medal Award in 1993, the Meteorological Society of Japan Fujiwara Award in 1994, and the Jule G. Charney Award from the American Meteorological Society in 1996. Kurihara has been a fellow of the American Meteorological Society since 1980.

Christopher W. Landsea (1965–)

Chris Landsea was born February 2, 1965, in Urbana, Illinois, but considers himself a native of south Florida, where he developed an interest in the hurricanes that are always a threat to this region. He first participated in hurricane research as a high school intern at the Hurricane Research Division, then attended the University of California in Los Angeles, where he earned a bachelor of science in atmospheric science in 1987. He then studied under the mentorship of William Gray, obtaining a master of science and Ph.D. in atmospheric science in 1991 and 1994, respectively. He then spent seven months as a Visiting Scientist at the Australian Bureau of Meteorology Research Centre in Melbourne and Macquarie University in Sydney to study the monsoon environment of hurricanes in that region. He is currently employed as a Research Meteorologist at the NOAA/Atlantic Oceanographic and Meteorological Laboratory's Hurricane Research Division.

Landsea's hurricane research determined the connection between African rainfall and Atlantic hurricane activity. He noted that the occurrence of major hurricanes (Category 3 or better) increases during rainy years in Africa. He also noted that the East Coast of the United States has experienced a downturn in major hurricane landfall during the 1970s to 1990s, a period that coincides with a 30-year drought in Africa. Landsea is a coauthor of Gray's annual hurricane seasonal forecast and is a prolific author of papers on seasonal and decadal hurricane cycles and hurricane climatology.

Landsea is the corecipient of the American Meteorological Society's Banner I. Miller Award, along with William M. Gray, Paul W. Mielke, Jr., and Kenneth J. Berry for the paper "Predicting Atlantic Seasonal Hurricane Activity 6–11 Months in Advance" at the May 1993 meeting of the Twentieth Conference on Hurricanes and Tropical Meteorology. The award was given for the "best contribution to the science of hurricane and tropical weather forecasting published during the years 1990–1992." He is also the recipient of the American Meteorological Society's Max A. Eaton Prize for the Best Student Paper, given at the Nineteenth Conference on Hurricanes and Tropical Meteorology in May 1991. He is on the 1997–2000 AMS Committee for Hurricanes and Tropical Meteorology and is an associate editor for the AMS journal *Weather and Forecasting*. The media frequently interviews Landsea about hurricanes.

Charles Neumann (1925–)

Charles Neumann participated in some of the earliest typhoon reconnaissance flights and is the primary developer of statistical hurricane track forecasts still being used today. Neumann was born in 1925 in New York City. In a World War II extension of the Naval ROTC program, he spent his freshman year at Holy Cross College in 1943, then transferred to the Massachusetts Institute of Technology, where he earned a bachelor of science in meteorology in 1946. Upon graduation, he volunteered as a U.S. Naval flight meteorologist for the first organized typhoon reconnaissance missions in the western North Pacific and served in that capacity in 1946 and 1947. Upon discharge from the Navy in 1947, he entered graduate school at the University of Chicago, where he earned a master of science degree in meteorology in 1949. From 1950 to 1952, Neumann was a civilian research meteorologist at the Air Weather Service Scientific Services Directorate in Washington, D.C. In 1952, he resumed his Naval career, having been recalled to active duty during the Korean Conflict and again served as a flight meteorologist—this time for Atlantic hurricanes. After one such flight in September 1953, which terminated at a Caribbean Island, he became infected with polio and retired from the Navy in 1954.

After two years of partial recovery from that illness, Neumann became involved in operational forecasting, serving as chief forecaster at Homestead Air Force Base (1956–1962) and aviation forecaster in Miami (1962–1966). During 1966–1971, he was a meteorologist with the NOAA Spaceflight Meteorology Group, which was collocated with the National Hurricane Center and provided forecasts for the Gemini and Apollo programs. In 1971, he joined the newly formed National Hurricane Center Research and Development Unit and became its chief in 1976, a position he held until his NOAA retirement in 1987. Since 1987, Neumann has been semiretired but works part-time as a senior scientist for Science Applications International Corporation (SAIC) providing hurricane risk information to military and insurance interests. He is currently working as a team member in revising the Atlantic hurricane historical database and will apply these data to a revised tropical cyclone risk model for the Atlantic.

Neumann is the developer of numerous global statistical and statistical-dynamical models for the prediction of tropical cyclone motion and intensity. These models include the Atlantic NHC 72, NHC 73, NHC 90, SHIFOR, CLIPER, the Eastern Pacific PE90 as well as many versions of these models for other ocean basins. Many of these models are still being used today at various worldwide forecast centers, including the National Hurricane Center. He has also developed statistical models for the assessment of hurricane risk and hurricane strike probability.

Neumann has authored more than 100 papers, mainly on hurricanes, and is the recipient of several awards. In 1971 he shared the Department of Commerce Silver Medal with fellow forecaster John Hope for "highly competent skill and ingenuity in developing objective techniques for use in hurricane predictions." He is the recipient of the 1977 American Meteorological Society's Banner I. Miller Award for "best published paper in tropical meteorology," and NOAA awarded NHC's Research and Development organization a citation for the "development of objective hurricane forecast aids in the eastern Pacific Ocean." In 1983, he was awarded the AMS Award for "Outstanding Contributions to Applied Meteorology." In 1986 Neumann received the Department of Commerce Gold Medal for "exceptional scientific achievement in statistical/dynamical track prediction modeling." Neumann is a fellow of the American Meteorological Society and has been a member of that organization for more than 50 years.

Grady Norton (1894–1954)

Grady Norton was an extraordinary hurricane forecaster in an era before computers and satellite provided weather guidance, and he was blessed with the ability to communicate with coastal residents during hurricane threats. Norton was born in 1894 in Womack Hill, Alabama, and became fascinated with severe weather as a young boy. He entered the Weather Bureau in 1915 and was drafted into the Army near the end of World War I. He served 10 months with the Signal Corps during 1918–1919 and attended a Corps' meteorology program at Texas A & M University.

Upon completing his military service, Norton returned to the Weather Bureau and worked at several different weather offices in the south central United States until 1935. Norton originally did not specialize in hurricane forecasting, but in 1928 he visited some relatives in south Florida in late summer and happened to drive into West Palm Beach during a mass funeral for the 1,836 victims of the Lake Okeechobee hurricane. Norton overheard a remark that such loss of life would not have occurred had adequate warnings been issued by the weather service. Although the comment was not entirely true, this event made a lasting impression on Norton, and he resolved to dedicate his life to the prevention of such tragedies.

In 1935, Congress appropriated funding for four hurricane forecast centers, one of which was established in Jacksonville, Florida. Norton was transferred there and named the office's chief hurricane forecaster. A few months later the Labor Day Hurricane hit the Florida Keys, killing 409 people even though Norton had issued warnings more than 12 hours in advance. This renewed Norton's resolve to make accurate forecasts and to better communicate warnings to the public. During the next 10 years, the average number of deaths from a major hurricane striking the United States was reduced from 50 to 5. Some of this success was due to remarkably accurate short-term track forecasts (considering the lack of observations at that time). However, Norton also effectively communicated warnings in a manner that the general public could understand. In February 1949 Norton received the Department of Commerce Silver Medal in recognition of his outstanding hurricane work. In 1943 the Jacksonville forecast center was moved to Miami to establish a joint hurricane warning service with the Air Corps and the Navy. Norton continued as the meteorologist in charge during this transition.

Norton suffered from high blood pressure that resulted in severe migraine headaches. However, he kept working against his doctor's advice. After a 12-hour stint forecasting for Hurricane Hazel on October 9, 1954, Norton suffered a stroke at home and passed away later that day. In 1955, the Miami office was officially designated as the National Hurricane Center, with Norton's understudy, Gordon Dunn, named as the first director. Most people, however, including Dunn, recognized Norton as the "honorary" first director of NHC.

Katsuyuki V. Ooyama (1929–)

Katsuyuki V. Ooyama (often referred to as "Vic" by colleagues) was born in Japan on March 5, 1929. He obtained a bachelor of science in physics in 1951 at the University of Tokyo. From 1951 to 1955, he worked at the Japan Meteorological Agency as a forecaster. During 1955, he enrolled at New York University, where he earned a master of science and Ph.D. under the mentorship of A. K. Blackadar and B. Haurwitz by studying atmospheric processes near the earth's surface. From 1958 to 1962, he participated in remote sensing of the ozone layer. From 1962 to 1973, he was a faculty member at New York University, then was a senior research scientist at the National Center for Atmospheric Research from 1973 to 1979. Since 1980 he has been a senior research scientist at the Atlantic Oceanographic and Meteorological Laboratory's Hurricane Research Division, developing three-dimensional weather models.

In 1969, Ooyama was the first scientist to realistically duplicate the basic two-dimensional features of a hurricane using a computer model, which he documented in the landmark paper "Numerical Simulation of the Life Cycle of Tropical Cyclones." In the same paper, Ooyama also performed the first multiscale perturbation analysis trying to explain how tropical genesis can occur based on the growth of small disturbances. This analysis launched a decade of theoretical research on tropical genesis and hurricane development based on similar mathematical theories. Ooyama is the recipient of the Meisinger Award (1968), the Fujiwhara Award (1971), and several Distinguished Authorship awards from the Environmental Research Laboratory of NOAA.

Henry Piddington (1797–1858)

Henry Piddington was the first person to coin the word "cyclone" for rotary storms (from the Greek word *kyklos,* meaning "coiling of the snake"). He was also a prolific writer about these weather systems, publishing a landmark book titled *The Sailor's Hornbook.* Born in Uckfield, East Sussex, England, Piddington began his career in the mercantile marines, eventually becoming a ship commander in the East India and China trade. In the late 1820s, Piddington became the curator of the Calcutta Museum, and during the next two decades he wrote articles for the Journal of the Asiatic Society on iron ore, minerals, soil used for produce, and plants. While in Calcutta, Piddington also became the president of the Marine Court and held the positions of officiating secretary, assistant secretary, and curator of the Geological Department of the Asiatic Society of Calcutta.

In the mid-1840s, Piddington became interested in rotary storms after reading William Reid's book *Law of Storms*. Recalling past experiences with such storms, especially one that shipwrecked Piddington's vessel and from which he narrowly escaped death, Piddington began collecting as many ship logs as possible that involved cyclones. Based on data from these logs, Piddington published 25 memoirs and a variety of books and journal articles. One book, *The Sailor's Hornbook*, includes documentation about previous researchers, such as William Redfield and William Reid. Piddington also performed research on ocean waves and the storm surge.

Piddington died in 1858.

William Redfield (1789–1857)

William Redfield made the first comprehensive analysis of the cyclonic rotation of hurricanes. Redfield was born on March 26, 1789, and after his seafarer father's death, he apprenticed as a saddle and harnessmaker in 1803. He completed his apprenticeship in 1810, and after some traveling in Ohio he made saddles and ran a store in Connecticut for a decade. Stimulated by an encounter with Fulton's steamboat during the Ohio trip, in 1822 Redfield began a career as a marine engineer and "transportation" promoter with a steamboat on the Connecticut River. He moved to New York in 1824 after the steamboat operations expanded to the Hudson River. In 1829, Redfield served as the superintendent of the steamboat company, in which he promoted a towboat operation for barges. Redfield was also involved in railroad promotion and planning.

Redfield was a self-taught scientist, with an insightful and remarkable organizational ability. While traveling around Connecticut after a strong hurricane in September 1821, he noted trees and corn in some areas were pointing toward the northwest, while those a few miles away were pointing to the southeast. He hypothesized that storms (and not just hurricanes) have a counterclockwise rotation, which he confirmed 10 years later when two more storms hit the New York area. A chance meeting with Professor Denison Olmsted of Yale, who thereafter constantly urged Redfield to print his theories, led Redfield to publish an article in the *American Journal of Science* in 1831 that concluded: "This storm exhibited in the form of a great whirlwind." This paper brought Redfield instant recognition, and sea captains provided Redfield with ship logs from which he published updated theories on storm rotation for the next 25 years in British and American journals. These later theories, explaining the source of rotation and storm development, were sometimes incorrect or incomplete and were the source of many lively discussions among scientists. However, he correctly noted the following observations about cyclones and hurricanes:

1. Winds do not rotate in horizontal circles but spiral inward toward the center
2. Cyclones rotate counterclockwise in the Northern Hemisphere and clockwise in the Southern Hemisphere, with wind speeds increasing toward the center
3. Cyclones move at a variable rate
4. Hurricanes generally form deep in the tropics, travel west-northwest with the tropical trade wind flow, and eventually turn back to the east as they propagate away from the tropics
5. Hurricanes are large, frequently with diameters greater than 1,000 miles
6. Pressure decreases with increasing rapidity near the center of a hurricane

Stimulated by his son's paper on fossil fish in 1836, Redfield also developed an interest in paleontology, publishing seven papers on the subject between 1838 and 1856 that established him as America's first specialist on fossil fish. He transformed the Association of American Geologists and Naturalists into the American Association for the Advancement of Science and was its first president in 1848. Yale University awarded Redfield an honorary degree in 1839; his name is commemorated by one genus and several species of fish and by Mount Redfield in the Adirondacks.

Redfield died on February 12, 1857.

Colonel William Reid (1791–1858)

William Reid was one of the first researchers of hurricanes and the author of a key book on the subject, titled *The Law of Storms.*

Born in Kinglassie, Reid was educated at the Military Academy at Woolwich, England. He entered the British army as an engineer in 1809, participated in many military engagements, and survived several conflict-induced injuries (including one diagnosed as fatal). His life was characterized by a lifetime of public service and scientific achievement.

After the wars, he was appointed major of engineers to restore government buildings after a hurricane hit Barbados in 1831. Upon seeing the destruction, Reid sought additional information about previous hurricanes in the hope of understanding them better. He collected detailed information about a number of hurricanes while in Barbados and developed theories on the circulation and movement of hurricanes. He spent the next few years writing papers about hurricanes and three editions of his book *The Law of Storms*. This book contained rules for mariners on how to avoid the most dangerous quadrants of hurricanes.

During 1838, Reid was appointed governor of Bermuda, and he aspired to improve the country's agriculture, education, and general welfare. In 1847 he established the first hurricane warning display system, to be used when the barometer indicated the approach of a hurricane. He returned to England in 1848 after the government refused to support the removal of a judge who oppressed the people. In 1849 he was appointed commanding engineer at Woolwich and was appointed governor of Malta in 1851 during the Russian war, in which he was promoted to general. In 1851 he also received the Order of Knighthood from the Queen.

Reid died October 31, 1858.

Lewis Fry Richardson (1881–1953)

Lewis Fry Richardson, a British mathematician and meteorologist, was born on October, 11, 1881, in Newcastle. He was educated at Durham College and Cambridge University, from which he graduated in 1903. In 1913 he became superintendent of the meteorological observatory at Eskdalemuir, Scotland. In 1920 he became head of the physics department at Westminster Training College. Nine years later he became the principal of Paisley Technical College, Scotland, where he remained until his retirement in 1940.

In 1922 Richardson published the classic 1922 book *Weather Prediction by Numerical Procedures*, which described how meteorological equations can be approximated to forecast the weather and also described the first attempt at "numerical forecasting." This effort was laborious because there were no electronic calculators or computers then, requiring many tedious hand calculations and a large support staff. In fact, Richardson estimated an operational numerical forecast for 2,000 locations would require 64,000 people just to carry out the calculations! Richardson attempted a 6-hour pressure forecast for just one location, and unfortunately it produced a grossly inaccurate prediction of 145 mb. It is now known that this failure was due to an improper approximation method and incorrect formulation of the equations. This forecast failure, along with the incredible amount of labor involved, discouraged further numerical modeling efforts for decades until computers were invented.

Nevertheless, Richardson's attempt at applying mathematics to weather forecasting was truly visionary. As computer speed increased, and as our understanding of the atmosphere and their governing equations improved, computer forecasts improved dramatically, eventually becoming the most important tool for hurricane forecasts (and weather forecasts in general). Richardson was also the first to envision a roomful of people performing mathematical calculations to predict the weather. Today, staff at the National Center for Environmental Prediction perform such a task on supercomputers.

Richardson worked on other aspects of meteorology, including turbulence studies. The "Richardson number," a fractional value that uses wind and temperature values to quantify the onset of turbulence, is named after him. Richardson also attempted to apply a mathematical framework to the causes of war, publishing several papers on the subject.

Richardson died September 30, 1953, at Kilmun, Scotland.

Herbert Riehl (1915–1997)

Widely regarded as the father of tropical meteorology, Herbert Riehl was born in Munich on March 30, 1915. He emigrated to the United States in 1933 and became a citizen in 1939. In 1942, he received a master of science degree in meteorology from New York University and taught in the Department of Meteorology at the University of Washington. When the U.S. Army Air Corps (Air Force) and the University of Chicago established the Institute of Tropical Meteorology on the campus of the University of Puerto

Rico in 1943 to provide weather research in equatorial regions, Riehl was a member of the team. He served as director of the institute from 1945 to 1946. He earned a Ph.D. from the University of Chicago in 1947 and remained on staff there until 1960, when he moved to Fort Collins, Colorado and established the Department of Atmospheric Science at Colorado State University. He served as department head until 1968 and as a professor until 1972. Even after Riehl left, Colorado State continued its reputation as a premier research institute on tropical meteorology.

In 1972, he accepted a post as director of the Institute of Meteorology and Geophysics at the Free University of Berlin. In 1976, he moved to Boulder, Colorado to join the staffs of the National Center for Atmospheric Research (NCAR) and the Cooperative Institute for Research in Environmental Sciences (CIRES), a joint laboratory of the federal government and the University of Colorado, as senior scientist. He held that post until 1989. After retirement, Riehl remained active in tropical meteorology research as a consultant in both North and South America.

Riehl profoundly influenced tropical meteorology with landmark papers and books on hurricane genesis, motion, intensification, wind shear interaction, and energetics. During his career he received many honors, including the AMS Meisinger Award in 1948, the Losey Award of the American Institute for Aeronautics and Astronautics in 1962, and the Carl-Gustav Rossby Research Medal in 1979. In May 1997 Riehl's contributions to tropical meteorology were honored at the Twenty-second Conference on Hurricanes and Tropical Meteorology of the American Meteorological Society.

One of his passions was mountain climbing. He climbed many mountains in Colorado and the United States, and major peaks throughout the world, including Kilimanjaro and Mont Blanc. Riehl died on June 1, 1997.

Carl-Gustav Arvid Rossby (1898–1957)

Carl-Gustav Rossby was born on December 28, 1898, in Stockholm, Sweden and was educated at the University of Stockholm. In 1919 he joined the Geophysical Institute at Bergen, which at the time, under Vilhelm Bjerknes, was the world's main center for meteorological research. In 1925 Rossby emigrated to the United States.

Rossby was one of the most eminent meteorologists of this century. Although he never actually performed any hurricane research, his influence was so encompassing in meteorology that his accomplishments must at least be mentioned. He made major contributions to our understanding of air masses and air movements. In the late 1930s Rossby simplified the equations Richardson attempted to solve by noting that certain terms could be neglected for large-scale weather. From these simplified equations, he also derived an expression for large undulations in the upper atmosphere now known as Rossby Waves. Because Rossby's research laid the foundation for many future scientific advances, he is known as the father of modern meteorology.

Rossby originally worked in Washington, D.C., then became a professor at the Massachusetts Institute of Technology. He left to found a meteorology program at the University of Chicago. He returned to Sweden in 1948 and founded the Institute of Meteorology.

Rossby died on August 19, 1957, in Stockholm, Sweden.

Joanne Simpson (1923–)

Joanne Simpson was the first woman meteorologist to earn a Ph.D., and she has made many contributions to the study of cloud physics and hurricanes. The daughter of an editor of the Boston Herald who reported on aviation as a hobby, Simpson was fascinated by flying and earned her student pilot's license at the age of 16. Because flying is weather-dependent, Simpson subsequently developed an interest in meteorology. She earned a bachelor of science degree in meteorology in 1943 at the University of Chicago. She then taught meteorology to aviation cadets and military forecasters while she pursued her master of science degree, which she completed in 1945. Unable to obtain a fellowship for a Ph.D. in an era when women were strongly discouraged from such aspirations, Simpson became a physics and meteorology instructor at the Illinois Institute of Technology. In 1947, Herbert Riehl lectured on aircraft observations of the wind flow and cloud structure in the tropics. Fascinated by this new field of tropical meteorology, she ended up completing her Ph.D. work in 1949 with Riehl as her adviser.

Riehl and Simpson wrote several landmark papers about hurricane structure, hurricane energetics, the thermodynamic structure of the tropics, and key concepts about the role of the tropical general circulation. From 1951 to 1960, Simpson became a research meteorologist at the Woods Hole Oceanographic Institute to learn more about weather over the ocean. During this period she constructed some of the first mathematical models of clouds and flew into clouds to validate her computations. In 1954 she won a Guggenheim Fellowship to work in England, and in 1955 she was an honorary lecturer at the Imperial College in London.

In 1962, Simpson wrote a chapter in the landmark book *The Sea, Volume I*. This book presented the most comprehensive treatment to date of the coupling between the ocean and the atmosphere. Simpson's chapter dealt with the complexity of how large- and small-scale atmospheric features interact, and in particular how difficult it would be to explicitly compute the individual contributions of small-scale clouds and turbulence on large-scale weather patterns. These complex scale interactions were factors that lead to the concept of "parameterization," in which the net effect of nonmeasurable small-scale weather features is computed in terms of large-scale (measurable) weather variables.

After five major hurricanes made landfall on the eastern United States in 1954 and 1955, Congress established the National Hurricane Research Project and named Simpson as an advisor. There she met the first director of the project, Robert Simpson (see next section), whom she married in 1965. From 1965 to 1979 she was director of NOAA's Experimental Meteorology Laboratory in Coral Gables, Florida, and participated in attempts to modify clouds and hurricanes using cloud-seeding techniques (see Chapter 1 about Project STORMFURY). In 1979 she went to the NASA Goddard Space Flight Center as head of the Severe Storm Branch; she has remained at Goddard ever since. She served as the project scientist for the Tropical Rainfall Measuring Mission (TRMM) from 1986 until its satellite launch in 1997 and is now chief scientist for meteorology. Simpson has won numerous awards for her achievements, including the American Meteorological Society's highest honor, the Carl-Gustav Rossby Research Medal.

Robert Simpson (1912–)

Robert Simpson's experience with hurricanes began when he was six years old in Corpus Christi, Texas, and a Category 4 hurricane hit the region in 1919. Simpson and his family waded through the rapidly rising water to the courthouse for shelter, where they then watched the hurricane decimate Corpus Christi. This experience forged his 34-year career as a hurricane scientist, through which he has made many contributions.

Simpson has participated in hundreds of reconnaissance flights into hurricanes, including some of the first Weather Bureau flights. Based on a 1946 flight, Simpson published a paper documenting the vertical structure of the hurricane eye. He also published several observational and theoretical papers on the eye and eyewall thermodynamic properties. Simpson became the first director of the National Hurricane Research Project (now the Hurricane Research Division) in 1955.

In 1959, Simpson went to the University of Chicago to finish his Ph.D., which had been interrupted by World War II. One day his advisor, Herbert Riehl, obtained permission from the Navy to accompany a flight through Hurricane Donna in 1960. Riehl noticed during this flight that the jet experienced icing, and the idea that perhaps a hurricane's structure and development could be modified with artificial ice nuclei was born. This idea ultimately culminated in the hurricane modification program known as Project STORMFURY (see Chapter 1 for details), for which Simpson was the chief proponent and first director. Project STORM-FURY was run at the National Hurricane Research Project laboratory, where he first met the lab's director, Joanne Malkus, whom he married in 1965 (see entry on Joanne Simpson).

In 1968, Simpson became director of the National Hurricane Center, where he remained until 1973. Simpson established a small research and development unit at NHC that developed many forecast applications. Simpson created the "hurricane specialist" position, which has forecast responsibilities during the hurricane season and research and public service in the remainder of the year. As NHC director, Simpson collaborated with a consulting engineer (Herbert Saffir) to devise a scale to quantify hurricane damage based on five intensity categories, known as the "Saffir-Simpson scale." In 1974, Simpson founded Simpson Weather Associates, where he continued contract work on matters related to hurricanes. In 1981, he coauthored with Herbert Riehl the important book *The Hurricane and Its Impact* (see Chapter 6).

Chris Velden (1956–)

Chris Velden has been a leader in the development of satellite applications to meteorology, with his primary focus on hurricane forecasting. Velden received his bachelor of science in 1979 from the University of Wisconsin-Stevens Point in natural science, with a minor in physics. He then earned a master of science degree in meteorology at the University of Wisconsin-Madison in 1982 with his study of satellite images of hurricanes in the microwave spectrum. For much of his career afterward he has remained at the University of Wisconsin to pursue satellite research, except for a visiting scientist position at the Australian Bureau of Meteorology in 1987–1988. Currently Velden is the Physical Science Senior Researcher at the Cooperative Institute for Meteorological Satellite Studies in the Space Science and Engineering Center located at the University of Wisconsin-Madison.

Velden and his colleagues have developed and automated a methodology for computing wind speed and direction by tracking individual cloud elements from satellite imagery. This procedure produces wind information in data-sparse regions, which is invaluable in tropical oceans. These winds are used to identify steering currents and to visualize atmospheric features that affect tropical cyclone genesis and intensification. These data are being assimilated in computer models, resulting in improved weather forecasts, including hurricane track predictions. For this effort, he received an award from the AMS in 1997. Velden and colleagues are also developing an objective satellite scheme that estimates a tropical cyclone's intensity in the hope it will supplement the current subjective satellite scheme known as the Dvorak technique. Velden has participated in numerous field experiments and has published over 20 articles.

Father Benito Viñes (1837–1893)

In his day, Father Benito Viñes was the greatest authority on hurricanes. Also known as the "hurricane priest," Viñes contributed to both theory and forecasting techniques and laid the foundation for today's observational and hurricane warning network. Viñes was born September 19, 1837, in Poboleda, Spain. With training in physics, Viñes came to Cuba from France in 1870 to serve as director of the Magnetic and Meteorological Observatory of the Royal College of Belen in Havana, Cuba, which had been established by Jesuit priests in 1857. After arriving, Viñes investigated ways to help people prepare for hurricanes by studying 12 years of records at the observatory, old newspaper accounts, and eyewitness accounts. His documentation was detailed and exhaustive, with hourly weather observations from 4 A.M. to 10 P.M. every day of pressure, temperature, moisture, and cloud heights. He also measured wind direction and speed at the surface and inferred wind aloft from cloud motion. After hurricane disasters, Viñes rummaged through debris and questioned survivors, meticulously documenting everything.

Based on this work, Viñes was the first to develop a methodology for locating a hurricane's center based on cloud motion aloft and sea swells. Implicit in this work was a remarkably accurate deduction of the hurricane's three-dimensional wind flow. He was also the first to forecast hurricane movement based on the motion of high clouds that flow out from the storm center. He also developed a methodology for predicting when a hurricane will recurve to the north, and for predicting average storm motion based on the latitude and time of year. This effort was augmented through the development of an observational network consisting of hundreds of observers along the Cuban coastline so that routine hurricane warnings could be issued. Routine forecasts began in 1875, with a pony express system distributing observations and forecasts between villages.

Viñes issued his first printed forecast on September 11, 1875, two days before an intense hurricane struck southern Cuba. Many lives were saved, and a legend was born. On September 14, 1876, Viñes warned of a dangerous hurricane offshore, and the captain of the ship *Liberty* ignored him, only to sail directly into the path of the storm and sink. In September 1877 Viñes accurately predicted that a hurricane would pass south of Puerto Rico but hit Santiago de Cuba. The same year Viñes published his most famous work, *Practical Hints in Regard to West Indian Hurricanes*, which quickly became standard learning material for mariners.

The revered Father Viñes died on July 23, 1893, in Havana, Cuba, three days after he mailed to the U.S. Weather Bureau another exceptional work about the three-dimensional wind flow of hurricanes titled *"Cyclonic Circulation and the Translatory Movement of West Indian Hurricanes."*

Hugh E. Willoughby (1945–)

Hugh Willoughby's impact on hurricane research consists of an interesting combination of high-level theory and observational analysis. Willoughby obtained a bachelor of science in geophysics geochemistry from the University of Arizona in 1967, followed by a master of science in 1969 from the Naval Postgraduate School. As a commissioned Navy officer, he served as a flight meteorologist in the Airborne Early Warning Squadron ONE from 1970 to 1971. From 1971 to 1974, Willoughby was a faculty member of the Naval Academy, where he taught meteorology, oceanography, geology, and computer science to midshipmen. Willoughby left active duty as a lieutenant to pursue a Ph.D. in atmospheric science at the University of Miami, which he earned in 1977.

While completing the Ph.D., Willoughby started working at the Hurricane Research Division (HRD) in NOAA's Atlantic Oceanographic and Meteorological Laboratory in 1975. He uses the data collected on the HRD reconnaissance flights to study hurricane structure and to formulate theoretical ideas. He first observed that hurricanes go through a natural (but temporary) weakening process in which a new eyewall forms outside the original eyewall. The outer eyewall "chokes off" inflow to the inner eyewall, causing it to dissipate. The outer eyewall then propagates inward, replacing the original eyewall. This process, known as the concentric eyewall cycle, lasts 12 to 36 hours and is associated with temporary weakening of the hurricane. This discovery, which led to the conclusion that hurricanes experience internal processes that influence intensification, implied that any perceived man-made changes to hurricanes in Project STORMFURY may have been the result of natural internal evolution instead.

Willoughby has made more than 400 research and reconnaissance flights into the eyes of typhoons and hurricanes in his career. He has occupied the G. J. Haltiner Visiting Research Chair at the Naval Postgraduate School (January–July 1991), was a Visiting Research Scientist at the Bureau of Meteorology Research Centre in Melbourne, Australia (June–July 1988), and was a Visiting Lecturer at the Shanghai Typhoon Institute (December 1985). He has published dozens of articles on hurricanes and is a fellow of the American Meteorological Society. As of 1998, he is the director of the Hurricane Research Division.

References

Anthes, R., 1982. "Tropical Cyclones: Their Evolution, Structure, and Effects." *American Meteorological Monographs*, vol. 19. Boston: American Meteorological Society, 208 pp.

Anthes, R. A., 1997. *Meteorology.* Upper Saddle River, NJ: Prentice Hall, 214 pp.

Burpee, R. W., 1988. Grady Norton: Hurricane Forecaster and Communicator Extraordinaire. *Wea. Forecasting,* 3:1050–1058.

Burpee, R. W., 1995. Necrologies—Gordon E. Dunn. *Bull. Amer. Meteor. Soc.* 76:260–261.

Byers, H. R., 1960. Carl-Gustaf Arvid Rossby, in *Biographical Memoirs,* Vol. 34. Washington, DC: National Academy of Sciences, 249–270.

Cabbage, M., 1998. Hurricane Man. *Sun Sentinel,* May 31.

Cline, I. M., 1926. *Tropical Cyclones.* New York: The Macmillan Company, 301 pp.

Cline, I. M., 1951. *Storms, Floods, and Sunshine.* New Orleans, LA: Pelican Publishing, 352 pp.

Daintith, J., S. Mitchell, E. Tootill, and D. Gjertsen, 1994. *Biographical Encyclopedia of Scientists,* Vol. 2. Aylesbury, UK: Market House Books Limited.

DeAngelis, D., 1989. The Hurricane Priest. *Weatherwise,* 42:256–257.

Elsberry, R. L., W. M. Frank, G. J. Holland, J. D. Jarrell, and R. L. Southern, 1987. *A Global View of Tropical Cyclones.* Monterey, CA: Naval Postgraduate School, 195 pp.

Foley, G. R., H. E. Willoughby, J. L. McBride, R. L. Elsberry, I. Ginis, and L. Chen, 1995. *Global Perspectives on Tropical Cyclones.* Geneva, Switzerland: World Meteorological Organization, Report No. TCP-38, 289 pp.

Gillispie, C. C., 1975. *Dictionary of Scientific Biography,* Vol. 11. New York: Charles Scriber's Sons.

Gray, W. M., C. W. Landsea, P. W. Mielke, and K. J. Berry, 1992. Predicting Atlantic Seasonal Hurricane Activity 6–11 Months in Advance. *Wea. Forecasting,* 7, 440–455.

Klinkenberg, J., 1999. "Hurricane Bill." *Floridian,* May 30, 1999.

Kutzbach, G., 1996. *The Thermal Theory of Cyclones: A History of Meteorological Thought in the Nineteenth Century.* Boston: American Meteorological Society, 272 pp.

Larson, E. 1999. *Isaac's Storm.* New York: Crown Publishers, 323 pp.

Malkus, J., 1962. Large Scale Interactions. In *The Sea: Ideas and Observations on Progress in the Study of the Seas,* Vol. 1. New York: Interscience Publishers. Chapter 4. Edited by M. N. Hill.

Ooyama, K., 1969. Simulation of the Life Cycle of Tropical Cyclones. *J. Atmos. Sci.,* 26:3–40.

Piddington, H., 1864. *The Sailor's Hornbook for the Law of Storms.* London: Williams and Norgate, 408 pp.

Redfield, W. C., 1831. Remarks on the Prevailing Storms of the Atlantic Coast, of the North American States. *American Journal of Science,* 20:17–51.

Reid, W., Sir, 1850. *An Attempt to Develop the Law of Storms by Means of Facts.* London: J. Weale, 530 pp.

Richardson, L. F., 1965. *Weather Prediction by Numerical Process (with a new introduction by Sidney Chapman).* New York: Dover Publications, 236 pp.

Simpson, R. H., and H. Riehl, 1981. *The Hurricane and Its Impact.* Baton Rouge, LA: Louisiana State University Press, 398 pp.

Viñes, B., 1885. *Practical Hints in Regard to West Indian Hurricanes.* Washington, DC: Weather Bureau, 15 pp.

Viñes, B., 1898. *Investigation of the Cyclonic Circulation and the Translatory Movement of West Indian Hurricanes.* Washington, DC: Weather Bureau, 34 pp.

Williams, J., 1992. *The Weather Book.* Arlington, VA: USA Today, 212 pp.

Williams, J., 1998. NASA Chief Kept Her Head in the Clouds. *USA Today,* Dec. 3.

Data, Opinions, Letters, and Mitigation 4

This chapter provides data, opinions, and letters relevant to hurricanes. Data are presented in tabular form concerning record events and annual figures since 1900. The opinions present a broad range of topics, including global warming, hurricane preparedness issues, and Project STORMFURY. Letters from survivors of hurricane landfalls in the 1700s and 1800s are included. The chapter closes with a discussion about hurricane mitigation and preparedness.

Data

Tables 4.1–4.9 show the deadliest, costliest, and most intense hurricanes in the United States, as well as the deaths, damages, hurricane progress, and hurricanes by category, in particular periods of time.

129

TABLE 4.1
The 30 Deadliest Landfalling Tropical Storms and Hurricanes in the U.S. Mainland, 1900–1998

Ranking	Hurricane	Year	Category	Deaths
1	Galveston, Texas	1900	4	8,000[a]
2	Lake Okeechobee, Florida	1928	4	1,836[a]
3	Florida Keys/S Texas	1919	4	600[b]
4	New England	1938	3[c]	600
5	Florida Keys	1935	5	408
6	Audrey—SW Louisiana/N Texas	1957	4	390
7	NE United States	1944	3[c]	390[d]
8	Grand Isle, Louisiana	1909	4	350
9	New Orleans, Louisiana	1915	4	275
10	Galveston, Texas	1915	4	275
11	Camille—Mississippi/Louisiana	1969	5	256
12	Miami and Pensacola, Florida/ Mississippi/Alabama	1926	4	243
13	Diane—NE United States	1955	1	184
14	SE Florida	1906	2	164
15	Mississippi/Alabama/Pensacola, Florida	1906	3	134
16	Agnes—NE United States	1972	1	125
17	Hazel—South Carolina/North Carolina	1954	4[c]	95
18	Betsy—SE Florida/SE Louisiana	1965	3	75
19	Carol—NE United States	1954	3[c]	60
20	SE Florida/Mississippi/Louisiana	1947	4	51
21	Donna—Florida/Eastern United States	1960	4	50
22	Georgia/South Carolina/ North Carolina	1940	2	50
23	Carla—Texas	1961	4	46
24	Velasco, Texas	1909	3	41
25	Freeport, Texas	1932	4	40
26	S Texas	1933	3	40
27	Hilda—Louisiana	1964	3	38
28	SW Louisiana	1918	3	34
29	Alberto—NW Florida/Georgia/ Alabama	1994	TS[e]	34
30	Amelia—S Texas	1978	TS[e]	33
Addendum (Pre-1900, California, and Caribbean Islands)				
	Louisiana	1893	Unknown	2,000
	South Carolina/Georgia	1893	Unknown	1,000–2,000
	Georgia/South Carolina	1881	Unknown	700
	San Felipe, Puerto Rico	1928	4	312
	U.S. Virgin Islands/Puerto Rico	1932	2	225
	Donna, St. Thomas, Virgin Islands	1960	4	107
	Southern California	1939	TS[e]	45
	Eloise, Puerto Rico	1975	TS[e]	44

(continues)

TABLE 4.1
(continued)

Ranking	Hurricane	Year	Category	Deaths
Addendum (Most Deadly Hurricanes in the Western Hemisphere)				
	The Great Hurricane, Martinique, St. Eustatius, Barbados, plus sinking of ships	1780	Unknown	22,000
	Mitch—Central America, including Honduras and Nicaragua	1998	1	11,000–18,000[f]
	Galveston, Texas	1900	4	8,000[a]
	Fifi—Honduras	1974	2	8,000
	Dominican Republic	1930	4	8,000
	Flora—Haiti, Cuba	1963	4 in Haiti, 3 in Cuba	7,200
	Point Petre Bay, Martinique	1776	Unknown	> 6,000

[a]May have actually been as high as 10,000 to 12,000
[b]Over 500 of these lost on ships at sea; 600–900 estimated deaths
[c]Moving more than 30 miles an hour
[d]Some 344 of these lost on ships at sea
[e]Only of tropical storm intensity
[f]Exact death count unknown
Unknown: Intensity not sufficiently known to establish category
Source: Hebert, P. J., J. D. Jarrell, and M. Mayfield, 1997. *The Deadliest, Costliest, and Most Intense United States Hurricanes of This Century (and Other Frequently Requested Hurricane Facts).* NOAA Technical Memorandum NWS TPC-1, National Oceanic and Atmospheric Administration, National Weather Service, National Hurricane Center, Miami, Florida, 30 pp. Also available at http://www.nhc.noaa.gov. Updated by the author through 1998. Additional information on Hurricane Mitch taken from the National Climatic Data Center's homepage http://www.ncdc.noaa.gov.

TABLE 4.2
The 30 Costliest Landfalling Tropical Storms and Hurricanes in the U.S. Mainland, 1900–1996

Ranking	Hurricane	Year	Category	Damage (in millions U.S. $)
1	Andrew—SE Florida/SE Louisiana	1992	4	26,500
2	Hugo—South Carolina	1989	4	7,000
3	Fran—North Carolina	1996	3	3,200
4	Opal—NW Florida/Alabama	1995	3	3,000
5	Frederic—Alabama/Mississippi	1979	3	2,300
6	Agnes—NE United States	1972	1	2,100
7	Alicia—N Texas	1983	3	2,000
8	Bob—North Carolina/NE United States	1991	2	1,500
9	Juan—Louisiana	1985	1	1,500
10	Camille—Mississippi/Alabama	1969	5	1,420.7
11	Betsy—Florida/Louisiana	1965	3	1,420.5
12	Elena—Mississippi/Alabama/ NW Florida	1985	3	1,250
13	Gloria—Eastern United States	1985	3[a]	900
14	Diane—NE United States	1955	1	831.7
15	Erin—Central and NW Florida/ SW Alabama	1995	2	700
16	Allison—N Texas	1989	T.S.[b]	500
16	Alberto—NW Florida/Georgia/ Alabama	1994	T.S.[b]	500
18	Eloise—NW Florida	1975	3	490
19	Carol—NE United States	1954	3[a]	461
20	Celia—S Texas	1970	3	453
21	Carla—Texas	1961	4	408
22	Claudette—N Texas	1979	T.S.[b]	400
23	Gordon—Central and S Florida/ North Carolina	1994	T.S.[b]	400
24	Donna—Florida/Eastern United States	1960	4	387
25	David—Florida/Eastern United States	1979	2	320
26	New England	1938	3[a]	306
27	Kate—Florida Keys/NW Florida	1985	2	300
27	Allen—S Texas	1980	3	300
29	Hazel—South Carolina/North Carolina	1954	4[a]	281
30	Bertha—North Carolina	1996	2	270

(continues)

TABLE 4.2
(continued)

Ranking	Hurricane	Year	Category	Damage (in millions U.S. $)
Addendum (non-Atlantic or non-Gulf Coast systems)				
8	Iniki—Kauai, Hawaii	1992	Unknown	1,800
8	Marilyn—U.S. Virgin Islands/ E Puerto Rico	1995	2	1,500
13	Hugo—U.S. Virgin Islands/Puerto Rico	1989	4	1,000
15	Hortense—Puerto Rico	1996	4	500
24	Olivia—California	1982	T.D.[c]	325
25	Iwa—Kauai, Hawaii	1982	Unknown	312
26	Norman—California	1978	T.D.[c]	300

Note: Costs are not adjusted for inflation.
[a]Moving more than 30 miles an hour
[b]Only of tropical storm intensity but included because of high damage
[c]Only a tropical depresssion
Unknown: Intensity not sufficiently known to establish category
Source: Hebert, P. J., J. D. Jarrell, and M. Mayfield, 1997. *The Deadliest, Costliest, and Most Intense United States Hurricanes of This Century (and Other Frequently Requested Hurricane Facts).* NOAA Technical Memorandum NWS TPC-1, National Oceanic and Atmospheric Administration, National Weather Service, National Hurricane Center, Miami, Florida, 30 pp. Also available at http://www.nhc.noaa.gov.

TABLE 4.3
The 30 Costliest Landfalling Hurricanes in the U.S. Mainland, 1900–1996, Inflation-Adjusted

Ranking	Hurricane	Year	Category	Damage (U.S. $)
1	Andrew–SE Florida/SE Louisiana	1992	4	30,475,000,000
2	Hugo–South Carolina	1989	4	8,491,561,181
3	Agnes–NE United States	1972	1	7,500,000,000
4	Betsy–Florida/Louisiana	1965	3	7,425,340,909
5	Camille–Mississippi/Alabama	1969	5	6,096,287,313
6	Diane–NE United States	1955	1	4,830,580,808
7	Frederic–Alabama/Mississippi	1979	3	4,328,968,903
8	New England	1938	3[a]	4,140,000,000
9	Fran–North Carolina	1996	3	3,200,000,000
10	Opal–NW Florida/Alabama	1995	3[a]	3,069,395,018
11	Alicia–N Texas	1983	3	2,983,138,781
12	Carol–NE United States	1954	3[a]	2,732,731,959
13	Carla–Texas	1961	4	2,223,696,682
14	Juan–Louisiana	1985	1	2,108,801,956
15	Donna–Florida/Eastern United States	1960	4	2,099,292,453
16	Celia–S Texas	1970	3	1,834,330,986
17	Elena–Mississippi/Alabama/ NW Florida	1985	3	1,757,334,963
18	Bob–North Carolina/NE United States	1991	2	1,747,720,365
19	Hazel–South Carolina/North Carolina	1954	4[a]	1,665,721,649
20	Miami, Florida	1926	4	1,515,294,118
21	Galveston and N Texas	1915	4	1,346,341,463[b]
22	Dora–NE Florida	1964	2	1,343,457,944
23	Eloise–NW Florida	1975	3	1,298,387,097
24	Gloria–Eastern United States	1985	3[a]	1,265,281,174
25	NE United States	1944	3[c]	1,064,814,815
26	Beulah–S Texas	1967	3	970,464,135
27	SE Florida/Louisiana/Mississippi	1947	4	810,897,436
28	Galveston, Texas	1900	4	809,207,317[d]
29	Audrey–Louisiana/N Texas	1957	4	802,325,581
30	Claudette–N Texas	1979	T.S.[e]	752,864,157
Addendum (non-mainland U.S. systems)				
16	Iniki–Kauai, Hawaii	1992	Unknown	2,070,000,000
20	Marilyn–U.S. Virgin Islands/ E Puerto Rico	1995	2	1,534,697,509
25	Hugo–U.S. Virgin Islands/Puerto Rico	1989	4	1,213,080,169
25	San Felipe, Puerto Rico	1928	4	1,150,000,000

Note: Costs are adjusted to 1996 dollars on basis of U.S. Department of Commerce Implicit Price Deflator for Construction. This does not include personal property increases and coastal county population changes.
[a]Moving more than 30 miles an hour
[b]Damage estimate was considered too high in 1915 reference
[c]Probably higher
[d]Using 1915 cost adjustment base; none available prior to 1915

[e]Only of tropical storm intensity but included because of high damage
Unknown: Intensity not sufficiently known to establish category

Source: Hebert, P. J., J. D. Jarrell, and M. Mayfield, 1997. *The Deadliest, Costliest, and Most Intense United States Hurricanes of This Century (and Other Frequently Requested Hurricane Facts)*. NOAA Technical Memorandum NWS TPC-1, National Oceanic and Atmospheric Administration, National Weather Service, National Hurricane Center, Miami, Florida, 30 pp. Updated to 1996 damage costs on http://www.nhc.noaa.gov.

TABLE 4.4
The 30 Costliest Landfalling Hurricanes in the U.S. Mainland, 1900–1995
(adjusted to 1995 dollars, for personal property increases, and for population changes)

Ranking	Hurricane	Year	Category	Damage (U.S. $)
1	SE Florida/Alabama	1926	4	72,303,000,000
2	Andrew—SE Florida/SE Louisiana	1992	4	33,094,000,000
3	Galveston, Texas*	1900	4	26,619,000,000
4	Galveston, Texas*	1915	4	22,602,000,000
5	SW Florida	1944	3	16,864,000,000
6	New England	1938	3	16,629,000,000
7	SE Florida/Lake Okeechobee	1928	4	13,795,000,000
8	Betsy—Florida/Louisiana	1965	3	12,434,000,000
9	Donna—Florida/Eastern United States	1960	4	12,048,000,000
10	Camille—Mississippi/Alabama	1969	5	10,965,000,000
11	Agnes—NE United States	1972	1	10,705,000,000
12	Diane—NE United States	1955	1	10,232,000,000
13	Hugo—South Carolina	1989	4	9,380,000,000
14	Carol—NE United States	1954	3	9,066,000,000
15	SE Florida/Louisiana/Alabama	1947	4	8,308,000,000
16	Carla—Central and N Texas	1961	4	7,069,000,000
17	Hazel—South Carolina/North Carolina	1954	4	7,039,000,000
18	NE United States	1944	3	6,536,000,000
19	SE Florida	1945	3	6,313,000,000
20	Frederic—Alabama/Mississippi	1979	3	6,293,000,000
21	SE Florida	1949	3	5,838,000,000
22	S Texas*	1919	4	5,368,000,000
23	Alicia—N Texas	1983	3	4,056,000,000
24	Celia—S Texas	1970	3	3,338,000,000
25	Dora—NE Florida	1964	2	3,108,000,000
26	Opal—NW Florida/Alabama	1995	3	3,000,000,000
27	Cleo—SE Florida	1964	2	2,435,000,000
28	Juan—Louisiana	1985	1	2,399,000,000
29	Audrey—Louisiana/N Texas	1957	4	2,396,000,000
30	King—SE Florida	1950	3	2,266,000,000

Note: Adjusted to 1995 dollars by inflation, personal property increases, and coastal county population changes (1900–1995). Asterisks indicate hurricanes included from years 1900–1924 using simplifying assumptions to extend the normalization methodology to 1900.
Source: Pielke, R. A., Jr., and C. W. Landsea, 1998. Normalized Atlantic Hurricane Damage: 1925–1995. *Wea. Forecasting,* 13:621–631.

TABLE 4.5
The Most Intense U.S. Hurricanes at Landfall, 1900–1998

Ranking	Hurricane	Year	Category (at landfall)	Pressure (in millibars)
1	Florida Keys	1935	5	892
2	Camille—Mississippi/SE Louisiana/ Virginia	1969	5	909
3	Andrew—SE Florida/SE Louisiana	1992	4	922
4	Florida Keys/S Texas	1919	4	927
5	Lake Okeechobee, Florida	1928	4	929
6	Donna—Florida/Eastern United States	1960	4	930
7	Galveston—Texas	1900	4	931
7	Grand Isle, Louisiana	1909	4	931
7	New Orleans, Louisiana	1915	4	931
7	Carla—Central and N Texas	1961	4	931
11	Hugo—South Carolina	1989	4	934
12	Miami and Pensacola, Florida/ Mississippi/Alabama	1926	4	935
13	Hazel—South Carolina/North Carolina	1954	4[a]	938
14	SE Florida/SE Louisiana/Mississippi	1947	4	940
15	N Texas	1932	4	941
16	Gloria—Eastern United States	1985	3[a, b]	942
16	Opal—NW Florida/Alabama	1995	3[b]	942
18	Audrey—SW Louisiana/N Texas	1957	4[c]	945
18	Galveston, Texas	1915	4[c]	945
18	Celia—S Texas	1970	3	945
18	Allen—S Texas	1980	3	945
22	New England	1938	3[a]	946
22	Frederic—Alabama/Mississippi	1979	3	946
24	NE United States	1944	3[a]	947
24	South Carolina/North Carolina	1906	3	947
26	Betsy—SE Florida/SE Louisiana	1965	3	948
26	SE Florida/NW Florida	1929	3	948
26	SE Florida	1933	3	948
26	S Texas	1916	3	948
26	Mississippi/Alabama	1916	3	948
31	Diane—North Carolina	1955	3[d]	949
31	S Texas	1933	3	949
33	Beulah—S Texas	1967	3	950
33	Hilda—Central Louisiana	1964	3	950
33	Gracie—South Carolina	1959	3	950
33	Texas Central	1942	3	950
37	SE Florida	1945	3	951
38	Tampa Bay, Florida	1921	3	952
38	Carmen—Central Louisiana	1974	3	952
40	Edna—New England	1954	3[a]	954
40	SE FLorida	1949	3	954

(continues)

TABLE 4.5
(continued)

Ranking	Hurricane	Year	Category (at landfall)	Pressure (in millibars)
40	Fran—North Carolina	1996	3	954
43	Eloise—NW Florida	1975	3	955
43	King—SE Florida	1950	3	955
43	Central Louisiana	1926	3	955
43	SW Louisiana	1918	3	955
43	SW Florida	1910	3	955
48	North Carolina	1933	3	957
48	Florida Keys	1909	3	957
50	Easy—NW Florida	1950	3	958
50	N Texas	1941	3	958
50	NW Florida	1917	3	958
50	N Texas	1909	3	958
50	Mississippi/Alabama	1906	3	958
55	Elena—Mississippi/Alabama/ NW Florida	1985	3	959
56	Carol—NE United States	1954	3[a]	960
56	Ione—North Carolina	1955	3	960
56	Emily—North Carolina	1993	3	960
59	Alicia—N Texas	1983	3	962
59	Connie—North Carolina/Virginia	1955	3	962
59	SW Florida/NE Florida	1944	3	962
59	Central Louisiana	1934	3	962
63	SW Florida/NE Florida	1948	3	963
64	NW Florida	1936	3	964

Note: Hurricanes may have had greater intensities at other times before landfall.
[a]Moving more than 30 miles an hour
[b]Highest category justified by winds
[c]Classified Category 4 because of estimated winds
[d]Cape Fear, North Carolina, area only; was a Category 2 at final landfall
Source: Hebert, P. J., J. D. Jarrell, and M. Mayfield, 1997. *The Deadliest, Costliest, and Most Intense United States Hurricanes of This Century (and Other Frequently Requested Hurricane Facts).* NOAA Technical Memorandum NWS TPC-1, National Oceanic and Atmospheric Administration, National Weather Service, National Hurricane Center, Miami, Florida, 30 pp. Also available at http://www.nhc.noaa.gov. Updated by the author through 1998.

TABLE 4.6
Estimated Annual Deaths and Damages from Hurricanes, 1900–1995

Year	Deaths	Damage ($ millions) Unadjusted	Adjusted	Year	Deaths	Damage ($ millions) Unadjusted	Adjusted
1900	8,000	30	790	1948	3	18	113
1901	10	1	26	1949	4	59	370
1902	0	Minor	Minor	1950	19	36	222
1903	15	1	26	1951	0	2	11
1904	5	2	53	1952	3	3	16
1905	0	Minor	Minor	1953	2	6	33
1906	298	3	79	1954	193	756	4,180
1907	0	0	0	1955	218	985	5,348
1908	0	0	0	1956	19	27	139
1909	406	8	211	1957	400	152	758
1910	30	1	26	1958	2	11	55
1911	17	1	26	1959	24	23	116
1912	1	Minor	Minor	1960	65	396	2,006
1913	5	3	79	1961	46	414	2,101
1914	0	0	0	1962	3	2	10
1915	550	63	1,660	1963	10	12	59
1916	107	33	723	1964	49	515	2,564
1917	5	Minor	Minor	1965	75	1,445	6,996
1918	34	5	71	1966	54	15	70
1919	287	22	278	1967	18	200	900
1920	2	3	30	1968	9	10	43
1921	6	3	38	1969	256	1,421	5,647
1922	0	0	0	1970	11	454	1,699
1923	0	Minor	Minor	1971	8	213	747
1924	2	Minor	Minor	1972	122	2,100	6,924
1925	6	Minor	Minor	1973	5	3	9
1926	269	112	1,415	1974	1	150	396
1927	0	0	0	1975	21	490	1,191
1928	1,836	25	315	1976	9	100	233
1929	3	1	12	1977	0	10	22
1930	0	Minor	Minor	1978	36	20	38
1931	0	0	0	1979	22	3,045	5,210
1932	0	0	0	1980	2	300	463
1933	63	47	701	1981	0	25	36
1934	17	5	68	1982	0	Minor	Minor
1935	414	12	163	1983	22	2,000	2,751
1936	9	2	28	1984	4	66	88
1937	0	Minor	Minor	1985	30	4,000	5,197
1938	600	306	3,864	1986	9	17	21
1939	3	Minor	Minor	1987	0	8	10
1940	51	5	66	1988	6	9	11
1941	10	8	98	1989	56	7,670	8,640

(continues)

TABLE 4.6
(continued)

Year	Deaths	Damage ($ millions)		Year	Deaths	Damage ($ millions)	
		Unadjusted	Adjusted			Unadjusted	Adjusted
1942	8	27	286	1990	16	57	63
1943	16	17	169	1991	16	1,500	1,637
1944	64	165	1,641	1992	24	26,500	28,687
1945	7	80	773	1993	4	57	59
1946	0	5	41	1994	38	973	973
1947	53	136	935	1995	29	3,723	3,582

Note: Both unadjusted and inflation-adjusted values are shown in the mainland United States from landfalling hurricanes and tropical storms during 1900–1995. The adjusted values are based on 1994 dollars using the U.S. Department of Commerce Implicit Price Deflator for Construction.

Source: Hebert, P. J., J. D. Jarrell, and M. Mayfield, 1997. *The Deadliest, Costliest, and Most Intense United States Hurricanes of This Century (and Other Frequently Requested Hurricane Facts).* NOAA Technical Memorandum NWS TPC-1, National Oceanic and Atmospheric Administration, National Weather Service, National Hurricane Center, Miami, Florida, 30 pp. Also available at http://www.nhc.noaa.gov.

TABLE 4.7
Progress of Average Atlantic Hurricane Season, Sorted by Date, 1944–1996

Number	Named Systems	Hurricanes	Category 3 or Greater
1	July 11	August 14	September 4
2	August 8	August 30	September 28
3	August 21	September 10	—
4	August 30	September 24	—
5	September 7	October 15	—
6	September 14	—	—
7	September 23	—	—
8	October 5	—	—
9	October 21	—	—

Note: Date upon which the following number of events would normally have occurred. This will exhibit considerable variability from year to year. However, in general most tropical storms and hurricanes occur in late August, September, and early October. Only occasionally will a major hurricane (Category 3 or better) occur in June or July.

Source: National Hurricane Center website at http://www.nhc.noaa.gov.

TABLE 4.8
Number of Landfalling U.S. Hurricanes by Category, 1900–1998

Category 5	2
Category 4	15
Category 3	47
Category 2	39
Category 1	59
Total	162
Major hurricanes (Categories 3, 4, 5)	64

Note: This shows that an average of two major hurricanes make landfall in the United States every three years, and that when all categories are combined an average of five hurricanes strike the United States every three years.
Source: Hebert, P. J., J. D. Jarrell, and M. Mayfield, 1997. *The Deadliest, Costliest, and Most Intense United States Hurricanes of This Century (and Other Frequently Requested Hurricane Facts).* NOAA Technical Memorandum NWS TPC-1, National Oceanic and Atmospheric Administration, National Weather Service, National Hurricane Center, Miami, Florida, 30 pp. Also available at http://www.nhc.noaa.gov. Updated by the author through 1998.

TABLE 4.9
List of Category 5 Atlantic Hurricanes (1886–Present)

Number	Storm Name	Maximum Wind	Date Attained (UTC)	Landfall as Category 5
1	Not Named	140 kt/160 mph	September 13, 1928	Puerto Rico
2	Not Named	140 kt/160 mph	September 5, 1932	Bahamas
3	Not Named	140 kt/160 mph	September 3, 1935	United States/Florida Keys
4	Not Named	140 kt/160 mph	September 19,1938	—
5	Not Named	140 kt/160 mph	September 16, 1947	Bahamas
6	Dog	160 kt/185 mph	September 6, 1950	—
7	Easy	140 kt/160 mph	September 7, 1951	—
8	Janet	150 kt/175 mph	September 28, 1955	Mexico
9	Cleo	140 kt/160 mph	August 16, 1958	—
10	Donna	140 kt/160 mph	September 4, 1960	—
11	Ethel	140 kt/160 mph	September 15, 1960	—
12	Carla	150 kt/175 mph	September 11, 1961	—
13	Hattie	140 kt/160 mph	October 30, 1961	—
14	Beulah	140 kt/160 mph	September 20, 1967	—
15	Camille	165 kt/190 mph	August 17, 1969	United States/Mississippi
16	Edith	140 kt/160 mph	September 9, 1971	Nicaragua
17	Anita	150 kt/175 mph	September 2, 1977	—
18	David	150 kt/175 mph	August 30, 1979	—
19	Allen	165 kt/190 mph	August 7, 1980	—
20	Gilbert	160 kt/185 mph	September 14, 1988	Mexico
21	Hugo	140 kt/160 mph	September 15, 1989	—
22	Mitch	155 kt/180 mph	October 26, 1998	—

Note: To qualify as a Category 5 hurricane on the Saffir-Simpson scale, maximum sustained winds must exceed 155 mph. Through 1998, only 22 Atlantic storms have reached this intensity, and only 8 were of category 5 strength at landfall. Of these 22, only 2 made U.S. landfall: the 1935 Florida Keys hurricane and Hurricane Camille, which hit the Mississippi coast in 1969. Above are listed all known Category 5 Atlantic hurricanes since records began in 1886.
Source: National Climatic Data Center at http://www.ncdc.noaa.gov.

Interesting Facts

Note that several infamous storms that struck the United States are listed in Table 4.9, but no entry appears in the "Landfall" column. These storms had weakened to below Category 5 intensity at the time of U.S. landfall. Hurricanes that had reached Category 5 intensity but had weakened by the time of U.S. landfall include hurricanes of 1928, 1938 (New England Hurricane), and 1947, plus Donna (1960), Ethel (1960), Carla (1961), Beulah (1967), David (1979), Allen (1980), and Hugo (1989). Other interesting facts include:

Most Intense Category 5 at U.S. Landfall, in Terms of Central Pressure.   1935 Florida Keys hurricane, with a central pressure of 892 mb (26.35 inches) and sustained wind speeds of 140 kts (160 mph).

Category 5 with Highest Winds at U.S. Landfall.   Hurricane Camille (1969), with a central pressure of 909 mb (26.84 inches) and sustained wind speeds of 165 kts (190 mph).

Most Intense Category 5 Atlantic Hurricane, in Terms of Central Pressure.   Hurricane Gilbert (1988), with a central pressure of 888 mb (26.22 inches) and sustained wind speeds of 160 kts (185 mph).

Longest Category 5*:   Hurricane Allen (1980), with a Central Pressure of 899 mb (26.55 inches) and Sustained Wind of 165 kts (190 mph).* Hurricane Allen reached Category 5 intensity three times along its path through the southern Caribbean and Gulf of Mexico: Twice these periods were of 24-hours duration and the third lasted 18 hours.

1.   With the exception of Camille, no Category 5 hurricanes have ever existed north of 30 degrees N or south of 14 degrees N.
2.   Four oceanic areas have experienced Category 5 intensity hurricanes twice: (26.5N, 77W), (18N, 86W), (24.5N, 96.5W) and (28–30N, 89W) (the path of Camille).
3.   Areas that have never experienced a landfalling hurricane of Category 5 intensity include: the U.S. East Coast, Cuba, Jamaica, and most of the Windward or Leeward Islands.

Source: National Climatic Data Center, http://www.ncdc. noaa.gov.

World Records

Most Intense Ever.   870 mb (over 180 mph) in Supertyphoon Tip, western North Pacific Ocean, October 12, 1979; reports from JTWC indicate that the 12 most intense hurricanes on record occurred in the western North Pacific Ocean.

Most Intense, Atlantic Ocean.   888 mb in Hurricane Gilbert, in September 1988. Although the pressure is almost 20 mb higher than western Pacific storms, the maximum sustained winds are comparable at over 180 mph.

Fastest Intensification.   100 mb (from 976 to 876 mb) in just under 24 hours in Supertyphoon Forrest, western North Pacific Ocean. Estimated surface winds increased from 75 mph to 173 mph during this period.

Extreme Storm Surge.   42 feet in the Bathurst Bay Hurricane, Northern Australia, 1899; this value was derived from reanalysis of debris sightings and eyewitness reports, and as a result is controversial. But clearly a phenomenal storm surge occurred.

Highest Wave.   112 ft from USS *Ramapo* in the western North Pacific, February 6–7, 1933; also in the western North Pacific, 82 ft on September 26, 1935.

Highest Rainfall.   All have occurred at La Reunion Island, Indian Ocean.

12 h: 45 in. at Foc-Foc (7,511-ft altitude) in Tropical Cyclone Denise, January 7–8, 1966

24 h: 71 in. at Foc-Foc in Tropical Cyclone Denise, January 7–8, 1966

48 h: 96 in. at Aurere (3,083-ft altitude) April 8–10, 1958

72 h: 126 in. at Grand-Ilet (3,773-ft altitude) in Tropical Cyclone Hyacinthe, January 18–27, 1980

10 days: 221 in. at Commerson (7,610-ft altitude) in Tropical Cyclone Hyacinthe, January 18–27, 1980.

Largest.   683-mile radius of gale force winds (34 mph) in Supertyphoon Tip, western North Pacific Ocean, October 12, 1979.

Smallest.   31-mile radius of gale force winds (34 mph) for Tropical Cyclone Tracy, Darwin, Australia, December 24, 1974. A storm this small is called a "midget" hurricane.

Smallest Eye.   4-mile radius from radar in Tropical Cyclone Tracy, Darwin, Australia, December 24, 1974.

Largest Eye.   56-mile radius from aircraft reconnaissance, Tropical Cyclone Kerry, Coral Sea, February 21, 1979.

TABLE 4.10
Record Number of Storms by Ocean Basin

Basin	Total Storms (tropical storms or greater with sustained winds of 39 mph or more)			Hurricane/Typhoon/Severe Tropical Cyclone (with sustained winds of 73 mph or more)		
	Most	Least	Average	Most	Least	Average
Atlantic	19	4	9.7	12	2	5.4
Northeast Pacific	23	8	16.5	14	4	8.9
Northwest Pacific	35	19	25.7	24	11	16.0
North Indian	10	1	5.4	6	0	2.5
Southwest Indian	15	6	10.4	10	0	4.4
Australian/Southeast Indian	11	1	6.9	7	0	3.4
Australian/Southwest Pacific	16	2	9.0	11	2	4.3
Globally	103	75	83.7	65	34	44.9

Note: Based on data from 1968–1989 for all Northern Hemisphere ocean basins except the Atlantic Ocean, and from 1968/69 to 1989/90 for the Southern Hemisphere. Starting in 1944, systematic aircraft reconnaissance was commenced for monitoring tropical storms and hurricanes in the Atlantic, hence these records are valid since 1944 in this basin. Atlantic records also contain a storm with some characteristics similar to hurricanes called subtropical cyclones.

Source: Landsea, C., 1999. Frequently Asked Questions: Hurricanes, Typhoons, and Tropical Cyclones. Available at http://www.aoml.noaa.gov/hrd/tcfaq.

TABLE 4.11

Maximum and Minimum Number of Atlantic Storms, and Maximum and Minimum United States Landfalls

Category	Maximum	Minimum
Tropical storms/hurricanes	19* (1995)	4 (1983)
Hurricanes	12 (1969)	2 (1982)
Major hurricanes (Category 3 or more)	7 (1950)	0 (many times, 1994 last)
U.S. landfalling tropical storms/hurricanes	8 (1916)	1 (many times, 1991 last)
U.S. landfalling hurricanes	6 (1916, 1985)	0 (many times, 1994 last)
U.S. landfalling major hurricanes	3 (1909, 1933, 1954)	0 (many times, 1994 last)

Note: The maximum and minimum number of storms in the Atlantic between 1944 and 1995, and the maximum and minimum number of storms making landfall between 1899 and 1995. Storm records are reliable beginning in 1944, when systematic reconnaissance flights began. Because of highly populated coastlines along the East Coast and Gulf Coast of the United States, reliable data extends back to 1899 for landfalling storms.

*1933 is recorded as being the most active of any Atlantic basin season on record (reliable or otherwise) with 21 tropical storms and hurricanes. However, since reconnaisance flights were not conducted at that time, it is excluded from this table.

Source: Landsea, C., 1998: Frequently Asked Questions: Hurricanes, Typhoons, and Tropical Cyclones (available at http://www.aoml.noaa.gov/hrd/tcfaq/).

Longest Lived.   Hurricane/Typhoon John lasted 31 days as it traveled across both the Northeast and Northwest Pacific Oceans during August and September 1994. In the Atlantic, the longest lived was Hurricane Ginger, which lasted 28 days in 1971.

Most Deaths.   As quoted from Holland (1993), "The death toll in the infamous Bangladesh Cyclone of 1970 has had several estimates, some wildly speculative, but it seems certain that at least 300,000 people died from the associated storm surge in the low-lying deltas." Also, several disastrous floods from landfalling typhoons in the Yangtze River Valley occurred in the mid-1850's resulted in many millions of deaths.

Costliest.   Hurricane Andrew (1992) in Florida and Louisiana is estimated to have caused at least $25 billion.

Source: Holland, G. J., 1993. In: *Global Guide to Tropical Cyclone Forecasting.* World Meteorological Organization Technical Document, WMO/TD No. 560, Tropical Cyclone Programme, Report No. TCP-31, Geneva, Switzerland. Also available on the web at http://www.bom.gov.au/bmrc (look for additional links under the mesoscale section to find this book).

Landsea, C., 1998: Frequently Asked Questions: Hurricanes, Typhoons, and Tropical Cyclones (available at http://www.aoml.noaa.gov/hrd/tcfaq/).

TABLE 4.12
Number of Landfalling Hurricanes for Each State during 1900–1998

Area	Category Number					All (1, 2, 3, 4, 5)	Major (3, 4, 5)
	1	2	3	4	5		
U.S. (Texas to Maine)	59	39	47	15	2	162	64
Texas	12	9	9	6	0	36	15
North	7	3	3	4	0	17	7
Central	2	2	1	1	0	6	2
South	3	4	5	1	0	13	6
Louisiana	9	5	8	3	1	26	12
Mississippi	1	2	5	0	1	9	6
Alabama	5	1	5	0	0	11	5
Florida	17	16	17	6	1	57	24
Northwest	10	8	7	0	0	25	7
Northeast	2	7	0	0	0	9	0
Southwest	6	4	6	2	1	19	9
Southeast	5	10	7	4	0	26	11
Georgia	1	4	0	0	0	5	0
South Carolina	6	4	2	2	0	14	4
North Carolina	10	5	10	1[a]	0	26	11
Virginia	2	1	1[a]	0	0	4	1[a]
Maryland	0	1[a]	0	0	0	1[a]	0
Delaware	0	0	0	0	0	0	0
New Jersey	1[a]	0	0	0	0	1[a]	0
New York	3	1[a]	5[a]	0	0	9	5[a]
Connecticut	2	3[a]	3[a]	0	0	8	3[a]
Rhode Island	0	2[a]	3[a]	0	0	5[a]	3[a]
Massachusetts	2	2[a]	2[a]	0	0	6	2[a]
New Hampshire	1[a]	1[a]	0	0	0	2[a]	0
Maine	5[a]	0	0	0	0	5	0

Note: Number of U.S. landfalling hurricanes for each state during 1900–1998, stratified by the Saffir-Simpson category. Some of these hurricanes may have hit land more than once, and each hit is counted individually. As a result, state totals will not equal U.S. totals, and Texas and Florida totals will not necessarily equal sum of sectional totals.
[a]All hurricanes in this group were moving faster than 30 mph
Source: Hebert, P. J., J. D. Jarrell, and M. Mayfield, 1997. *The Deadliest, Costliest, and Most Intense United States Hurricanes of This Century (and Other Frequently Requested Hurricane Facts).* NOAA Technical Memorandum NWS TPC-1, National Oceanic and Atmospheric Administration, National Weather Service, National Hurricane Center, Miami, Florida, 30 pp. Also available at http://www.nhc.noaa.gov. Updated by the author through 1998.

Hurricane Fatalities since 1970

Historically, most deaths associated with tropical storms and hurricanes were caused by the storm surge. As evacuation procedures have improved in the United States, fatalities caused by

TABLE 4.13
Deadliest Top 10 U.S. Tropical Storms and Hurricanes during 1970–1998

Rank	Name	Year	Dead
1	Agnes	1972	125
2	Alberto	1994	34
3	Amelia	1978	33
4	Celia	1970	25
5	Andrew	1992	23
6	Fran	1996	22
7	Hugo	1989	17
8	Alicia	1983	13
8	Chantal	1989	13
8	Charley	1998	13

Note: Deaths incurred from heart attacks, house fires, and vehicle accidents during the storms are considered "indirect" and are excluded from the database.

TABLE 4.14
Tropical Storm and Hurricane Deaths during 1970–1998

Cause	Percentage of Deaths	Number of Deaths
Drowned	81%	(415/510)
Wind[a]	13%	(66/510)
Tornado	4%	(20/510)
Other[b]	2%	(9/510)

Note: There are a total of 517 deaths, with 7 being caused by unknown circumstances.
[a]From building failures, airborne debris, etc.
[b]From lightning, hypothermia, downed aircraft

TABLE 4.15
Tropical Storm and Hurricane Deaths by Drowning during 1970–1998

Cause	Percentage of All Drowned	Percentage of All Deaths
Inland flooding	71% (292/411)	57% (292/510)
Shoreline[a]	15% (62/411)	12% (62/510)
Offshore[b]	13% (52/411)	10% (53/510)
Storm surge	1% (5/411)	1% (5/510)

Note: There are 415 drowning deaths, and 4 by unknown causes.
[a]Rough surf, rip currents, large swells
[b]Beyond breakers to 50 nautical miles from shore

TABLE 4.16
Location of U.S. Tropical Storm and Hurricane Deaths during 1970–1998

Location	Percentage of Deaths	Number of Deaths
Inland county	62%	(295/478)
Coastal county[a]	27%	(128/478)
Offshore[b]	12%	(55/478)

Note: There are 517 deaths, and 39 in uncertain locations.
[a]Includes storm surge, large swells, and rip current areas
[b]Beyond breakers to 50 nautical miles from shore
Source: Rappaport, E. N., M. Fuchs, and M. Lorentson, 1999. The threat to life in inland areas of the United States from Atlantic tropical cyclones. In *Preprint from the 23rd Conference on Hurricanes and Tropical Meteorology.* Boston: American Meteorological Society, 339–342.

the storm surge have diminished. In fact, most deaths in tropical storms and hurricanes are now associated with inland flooding, as shown in Tables 4.13, 4.14, and 4.15.

Opinions

Excerpts from "Project STORMFURY: A Scientific Chronicle 1962–1983"

This article chronicles Project STORMFURY, in which reconnaissance flights attempted to weaken hurricanes by "seeding" them with silver iodide. Please refer to Chapter 1 for a discussion of Project STORM-FURY. Willoughby et al. explains the motivation for Project STORM-FURY in the following excerpt.

The years 1954 and 1955 each brought three major hurricanes to the East Coast of the United States. Hurricanes Carol, Edna, Hazel, Connie, Diane, and Ione together destroyed more than six billion dollars in property (adjusted to 1983) and killed nearly 400 U.S. citizens. Six hurricane landfalls in two years seemed to call for some sort of governmental action. In the spring of 1955, the U.S. Congress appropriated substantially more money for hurricane research. After the 1955 hurricane season, it mandated that the U.S. Weather Bureau establish the National Hurricane Research Project (NHRP), which was to become the National Hurricane Laboratory in 1964 and the Hurricane Research Division of the Atlantic Oceanographic and Meteorological Laboratory in 1983. The mission of NHRP was fourfold: to study the formation of hurricanes; to study their structure and dynamics; *to seek means for hurricane modification;* and to seek means for improvement of forecasts. The third objective was, in time, to provide justification for Project STORMFURY."

For a variety of political and scientific reasons, funding for STORM-FURY was threatened in the late 1970s. Below Willoughby et al. list five criteria needed to justify the continuation of STORMFURY and then explain why the project was eventually abandoned.

STORMFURY had to be viable in five different respects [to justify continuation]:

> *Political.* Governments had to be willing to accept the risk of a public outcry if a seeded hurricane (or typhoon) devastated a coastal region. This outcry and its legal consequences might arise even if human intervention had no effect at all on the hurricane.
> *Operational.* The aircraft, instrumentation, and personnel to do the seeding and to document the result had to be available.
> *Microphysical.* Convection in hurricanes had to contain enough supercooled water for seeding to be effective.
> *Dynamic.* The hurricane vortex had to be sufficiently labile for human intervention to change its structure.
> *Statistical.* The experiment had to be repeatable, and the results had to be distinguishable from natural behavior.

STORMFURY itself, however, had two fatal flaws: it was neither microphysically nor statistically feasible. Observational evidence indicates that seeding in hurricanes would be ineffective because they contain too little supercooled water and too much natural ice. Moreover, the expected results of seeding are often indistinguishable from naturally occurring intensity changes. By mid-1983, none (or perhaps only one: dynamic feasibility) of the five conditions for development of an operational hurricane amelioration strategy could be met, and Project STORMFURY ended. Its lasting legacies are the instrumented aircraft and two decades of productive research. Of the four original objectives for hurricane research set down in 1955 at the establishment of NHRP, three—understanding formation, understanding structure and dynamics, and improvement of forecasts—remain areas of active investigation.

Source: Willoughby, H. E., D. P. Jorgensen, R. A. Black, and S. L. Rosenthal, 1985: Project STORMFURY: A Scientific Chronicle 1962–1983. *Bull. Am. Meteor. Soc.*, 66:505–514.

Excerpts from the 1996 IPCC Report on Whether Global Warming Is Influencing Global Hurricane Activity

The Intergovernmental Panel on Climate Change (IPCC) is a group of over 200 scientists that originally met in 1990 to discuss issues related to possible humanity-induced climate change. Based on these and subsequent meetings in 1992, 1994, and 1996, the IPCC issues conclusions regarding global warming based on current data and the state of the science. The 1996 IPCC report states the following:

The state-of-the-art for [hurricane simulations in numerical models] remains poor because: (i) tropical cyclones cannot be adequately resolved in general circulation models; (ii) some aspects of ENSO are not well-simulated in general circulation models; (iii) other large-scale changes in the atmospheric general circulation which could affect tropical cyclones cannot yet be discounted; and (iv) natural variability of tropical cyclones is very large, so small trends are likely to be lost in the noise. . . .In conclusion, it is not possible to say whether the frequency, area of occurrence, mean intensity, or maximum intensity of tropical cyclones will change.

Source: Houghton, J. T., L. G. Meira Filho, B. A. Callander, N. Harris, A. Kattenberg, and K. Maskell, 1996. *Climate Change 1995: The Science of Climate Change. Contribution of Working Group I to the Second Assessment of the Intergovernmental Panel on Climate Change.* Cambridge: Cambridge University Press, 572 pp.

Excerpt from "Tropical Cyclones and Global Climate Change: A Post-IPCC Assessment"

As a follow-up to the 1996 IPCC report, a group of tropical cyclone experts issued some more definitive statements regarding any relationships to possible global warming and tropical cyclone activity in 1998:

Our knowledge has advanced to permit the following summary.

- There are no discernible global trends in tropical cyclone number, intensity, or location from historical data analyses;
- Regional variability, which is very large, is being quantified slowly by a variety of methods;

- Empirical methods do not have skill when applied to tropical cyclones in greenhouse conditions;
- Global and mesoscale model-based predictions for tropical cyclones in greenhouse conditions have not yet demonstrated prediction skill;
- There is no evidence to suggest any major changes in the area or global location of tropical cyclone genesis in greenhouse conditions;
- Thermodynamic "upscaling" models seem to have some skill in predicting maximum potential intensity (MPI); and
- These thermodynamic schemes predict an increase in MPI of 10%–20% for a doubled CO_2 climate but the known omissions [from MPI theory] all act to reduce these increases.

Source: Henderson-Sellers, A., H. Zhang, G. Berz, K. Emanuel, W. Gray, C. Landsea, G. Holland, J. Lighthill, S-L. Shieh, P. Webster, and K. McGuffie, 1998. Tropical Cyclones and Global Climate Change: A Post-IPCC Assessment. *Bull. Am. Meteor. Soc.*, 79:19–38.

Excerpts from "Catastrophic Hurricanes May Become Frequent Events in Caribbean and along the United States East and Gulf Coast"

Hurricane Andrew remains the worst economic natural disaster inflicted on the United States. This article, written by then-NHC director Bob Sheets, succinctly discusses the economic losses, the hurricane's impact on insurance companies, the relative lack of funding compared to other natural hazard mitigation research, the hurricane preparedness problem, and the sobering prospects of an increase in hurricanes in the future. Although this article was written in 1993, its comments are very much relevant today.

Hurricane Andrew in 1992 resulted in the largest economic loss from a natural disaster in United States history. Total losses are estimated to be about $25 billion, which is more than the total of the three previous most costly hurricanes. Insured losses are now estimated at $18 billion, more than four times the previous record payout which occurred with Hurricane Hugo. State Farm

Insurance had the record for a payout by a single insurance company with nearly $500 million for Hurricane Hugo. It has been reported that State Farm has paid out more than $3.6 billion for losses that occurred in Hurricane Andrew, more than seven times the previous record. Eight insurance companies have failed and as reported by Florida's Insurance Commissioner, major problems are ahead in the insurance situation for all coastal areas subject to hurricanes. Many insurance companies have either withdrawn from coastal markets or are limiting market shares making private insurance difficult to obtain in many areas.

It Could Have Been Much Worse

Had Hurricane Andrew been displaced only 20 miles north of its westward track over South Florida, two different studies show that losses would have exceeded $75 billion in Southeast Florida alone. A continuation of that same track across Florida would have resulted in major losses in the Ft. Myers area and would have resulted in our nightmare storm into New Orleans. Total property damage would have been far in excess of $100 billion. It is reasonable to assume that the nation's insurance industry would have been in shambles and that losses from rising waters in Miami Beach and New Orleans would have far exceeded the reserves of the National Flood Insurance Program.

Of even greater concern is that casualties in the southeast Florida area would have been large because more than 30 percent of the people did not evacuate the condominium complexes on Miami Beach, Hallandale and Hollywood. Casualties in New Orleans could be very large with people caught in a grid lock on the streets which are below sea level as water topped the levee system, putting the city under 18 to 20 feet of water.

Evacuation Becoming More Difficult

It is becoming more difficult each year to evacuate people from barrier islands and other coastal areas due to roadway systems that have not kept pace with the rapid population growth. That is, in most coastal regions, it now takes much longer to evacuate a threatened area than it did a decade ago. Since Hurricanes Hugo and Andrew, many coastal residents who are out of the area which might be inundated by the storm surge now say they are going to evacuate because of damage caused by wind! This means that longer and longer lead times are being required for safe evacuation from threatened areas. Unfortunately, these required longer-range forecasts suffer from increased uncertainties.

A recent study in Southeast Florida indicated that it would take more than 80 hours to clear the area for those who say they will leave the next time a major hurricane threatens. Eighty hours before Andrew struck South Florida, it was just barely a tropical storm and almost equally likely to strike anywhere on the Florida East Coast. Providing such lead times is impractical with the present state of the science.

Potential for Large Loss of Life

The protection of life is the highest priority goal of the National Weather Service's Hurricane Forecast and Warning program and as such, is the primary factor that determines the degree of "over-warning." As much as 30 hours or more is now required to evacuate people from such vulnerable areas as Galveston Island, the Florida Keys, New Orleans, Ocean City, Maryland, etc. Such decisions for protection of life now must be based upon 36-hour or longer-range forecasts. The uncertainty in those forecasts requires relatively large "over-warning," in order to minimize the potential for large loss of life. Even with this "over-warning" there could be an unforeseen meteorological development (rapid change in course [Elena 1985], rapid change in intensity) which would not permit adequate warning time. Also, there could be some hindrance to the evacuation process (accidents, barge taking out a bridge, slow start of evacuation) which could also result in incomplete evacuations with literally thousands of people trapped on barrier islands and roadway systems as the life threatening elements (rising waters, increasing winds) of the hurricane approach.

Even support for hurricane preparedness and mitigation activities seems to be out of balance with expenditures for other natural hazards. It was reported at the 1993 National Hurricane Conference, that FEMA spends nearly $50 on earthquake related programs for every one dollar it spends upon hurricane programs, excluding relief efforts. This does not mean that the earthquake problem is any less important than it has been, but clearly the hurricane problem has not received the attention that it needs.

One reason that we have arrived at this situation is that we had been in a fortunate period of a very limited number of major hurricane strikes on our coasts during the past few decades before Hurricanes Hugo and Andrew. From 1941 through 1950, there were 10 major (Saffir/Simpson scale category 3 or stronger; sustained winds greater than 110 mph) hurricanes striking the continental United States. Seven of those struck Florida. From 1951 through 1960, there were 8 major hurricanes striking the United States with 7 of those making landfall on the East Coast. For the next 30 years, the only major hurricane to strike the Florida peninsula was Hurricane Betsy of 1965. Similarly during this period, no major hurricanes made landfall on the East Coast until the mid-1980s.

Recent research by Professor William Gray of Colorado State University indicates that this cycle of hurricane activity may be related to rainfall over the western Sahel area of Africa. It should be noted that rainfall in (the western Sahel), where "seedlings" for many hurricanes originate, was generally above normal from 1949 through 1967 and then has been generally below normal. The trend now seems to be toward more normal rainfall in the West Africa region. If that trend continues then a return to activity similar to the 1940's and 1950's may occur in the near future.

Source: Sheets, Robert C., 1993. Catastrophic Hurricanes May Become Frequent Events in Caribbean and along the United States East and Gulf Coast. Proceedings from *Hurricanes of 1992: Lessons Learned and Implications for the Future.* Miami, FL: American Society of Civil Engineers, 37–51.

Letters

Excerpt from Letter Written by Alexander Hamilton to His Father Following Two Hurricanes

It was hurricanes that indirectly brought Alexander Hamilton to America. Hamilton experienced two hurricanes on August 3 and September 3, 1772, on St. Croix in the West Indies. His description of this experience in a letter to his father so impressed local planters that they took up a collection to send him to King's College (now Columbia University) in New York. He was in America when the first rumblings of revolution occurred, and the rest is history.

Good God! What horror and destruction! It is impossible for me to describe it or for you to form any idea of it. It seemed as if a total dissolution of nature was taking place. The roaring of the sea and wind, fiery meteors flying about in the air, the prodigious glare of almost perpetual lightning, the crash of falling houses, and the earpiercing shrieks of the distressed were sufficient to strike astonishment into Angels. A great part of the buildings throughout the island are leveled to the ground; almost all the rest very much shattered, several persons killed and numbers utterly ruined—whole families roaming about the streets, unknowing where to find a place of shelter—the sick exposed to the keenness of water and air, without a bed to lie upon, or a dry covering to their bodies, and out harbors entirely bare. In a word, misery, in its most hideous shapes, spread over the whole face of the country.

Sources: Hughes, P., 1976. *American Weather Stories.* Washington, DC: U.S. Department of Commerce, 116 pp.

Hughes, P., 1987. Hurricanes Haunt Our History. *Weatherwise,* 40:134–140.

The Great Hurricane

One of the most destructive hurricanes of the 1700s struck the West Indies in October 1780, killing around 22,000 people. This storm, called The Great Hurricane by many in this century, also inflicted great losses on both the fleets of Britain and England, who were planning assaults on each other to claim the Antilles Islands.

The first news of the disaster was carried to England in a letter from Major General Vaughn, commander-in-chief of His Majesty's forces in the Leeward Island, dated October 30, 1780:

I am much concerned to inform your Lordship, that this island was almost entirely destroyed by a most violent hurricane, which began on Tuesday the 10th instant. And continued almost without intermission for nearly forty-eight hours. It is impossible for me to attempt a description of the storm; suffice it to say, that few families have escaped the general ruin, and I do not believe that 10 houses are saved in the whole island: scarce a house is standing in Bridgetown; whole families were buried in the ruins of their habitations; and many, in attempting to escape, were maimed and disabled: a general convulsion of nature seemed to take place, and an universal destruction ensued. The strongest colours could not paint to your Lordship the miseries of the inhabitants: on the one hand, the ground covered with mangled bodies of their friends and relations, and on the other, reputable families, wandering through the ruins, seeking for food and shelter: in short, imagination can form but a faint idea of the horrors of this dreadful scene.

Admiral Rodney was in New York during the hurricane, and when he visited Barbados he incorrectly concluded that the storm must have been accompanied by an earthquake to achieve such total destruction. Rodney writes:

The whole face of the country appears an entire ruin, and the most beautiful island in the world has the appearance of a country laid waste by fire, and sword, and appears to the imagination more dreadful than it is possible for me to find words to express.

Source: Ludlum, D. M., 1989. *Early American Hurricanes 1492–1870.* Boston, MA: American Meteorological Society, 198 pp.

Eyewitness Account of Hurricane Landfall on Florida Panhandle in 1843

The following is an excerpt from the September 15, 1843, Commercial Gazette describing the damage on Port Leon (about 30 miles south of today's city of Tallahassee near St. Marks):

Our city is in ruins! We have been visited by one of the most horrible storms that it ever before devolved upon us to chronicle. On Wednesday [13 September] about 11 o'clock A.M. the wind lulled and the tide fell, the weather still continued lowering. At 11 at night, the wind freshened, and the tide commenced flowing, and by 12 o'clock it blew a perfect hurricane, and the whole town was inundated. The gale continued with unabated violence until 2 o'clock, the water making a perfect breach ten feet deep over our town. The wind suddenly lulled for a few minutes, and then came from southwest with redoubled violence and blew until daylight.

Every warehouse in the town was laid flat with the ground except one, Messers Hamlin & Shell's, and a part of that also fell. Nearly every dwelling was thrown from its foundations, and many of them crushed to atoms. The loss of property is immense. Every inhabitant participated in the loss more or less. None have escaped, many with only the clothes they stand in. St. Marks suffered in like proportion with ourselves. But our losses are nothing compared with those at the lighthouse. Every building but the lighthouse gone—and dreadful to relate fourteen lives lost! And among them some of our most valued citizens. We cannot attempt to estimate the loss of each individual at this time, but shall reserve it until our feelings will better enable us to investigate it.

Source: Ludlum, D. M., 1989. *Early American Hurricanes 1492–1870.* Boston, MA: American Meteorological Society, 198 pp.

The Last Island Disaster

On August 10, 1856, a violent hurricane struck Isle Derniere (Cajun for Last Island), a pleasure resort southwest of New Orleans. The highest points were under five feet of water. The resort hotel and surrounding gambling establishments were destroyed, and the island was cut in half by the storm. Crew members on the ferry boat Star rode out the storm and rescued survivors seen floating among the debris. A total of about 40 people survived on the Star; the rest—hundreds of people— were killed. The following letter was sent to the Daily Picayune (the New Orleans newspaper) from a survivor of this event and was published August 14, 1856:

Dear Pic.—You may have heard ere this reaches you of the dreadful catastrophe which happened on Last Island on Sunday the 10th inst. As one of the sufferers it becomes my duty to chronicle one of the most melancholy events, which have ever occurred. On Saturday night, the 9th inst., a heavy northeast wind prevailed, which excited the fears of a storm in the minds of many; the wind increased gradually until about ten o'clock Sunday morning, when there existed no longer any doubt that we were threatened with imminent danger. From that time the wind blew a perfect hurricane; every house upon the island giving way, one after another, until nothing remained. At this moment everyone sought the most elevated point on the island, exerting themselves at the same time to avoid the fragments of buildings, which were scattered in every direction by the wind. Many persons were wounded; some mortally. The water at this time (about 2 o'clock P.M.) commenced rising so rapidly from the bay side, that there could no longer be any doubt that the island would be submerged. The scene at this moment forbids description. Men, women, and children were seen running in every direction, in search of some means of salvation. The violence of the wind, together with the rain, which fell like hail, and the sand blinded their eyes, prevented many from reaching the objects they had aimed at.

At about 4 o'clock, the Bay and Gulf currents met and the sea washed over the whole island. Those who were so fortunate as to find some object to cling to, were seen floating in all directions. Many of them, however, were separated from the straw to which they clung for life, and launched into eternity; others were washed away by the rapid current and drowned before they could reach their point of destination. Many were drowned from being stunned by scattered fragments of the buildings, which had been blown asunder by the storm; many others were crushed by floating timbers and logs, which were removed from the beach, and met them on their journey. To attempt a description of this sad event would be useless. No words could depict the awful scene which occurred on the night between the 10th and 11th inst. It was not until the next morning the 11th, that we could ascertain the extent of the disaster. Upon my return, after having drifted for about twenty hours, I found the steamer *Star*, which had arrived the day before, and was lying at anchor, a perfect wreck, nothing but her hull and boilers, and a portion of her machinery remaining. Upon this wreck the lives of a large number were saved. Toward her each one directed his path as he was recovered from the deep, and was welcomed with tears by his fellow-sufferers, who had been so fortunate as to escape. The scene was heart-rending; the good fortune of many an individual in being saved, was blighted by the news of the loss of a father, brother, sister, wife or some near relative.

Sources: Ludlum, D. M., 1989. *Early American Hurricanes 1492–1870.* Boston, MA: American Meteorological Society, 198 pp.

Jennings, G., 1970. *The Killer Storms: Hurricanes, Typhoons, and Tornadoes.* New York: J. B. Lippincott Co., 207 pp.

Mitigation

Excerpts from the FEMA Document "Against the Wind—Protecting Your Home from Hurricane Wind Damage"

This article discusses construction techniques that will strengthen the roof, doors, windows, and garage doors against hurricane-force winds. The interested reader is encouraged to order the document from FEMA, since it also contains drawings that clarify the instructions. The document is also available from FEMA's website at http://www.fema.gov.

Introduction

During a hurricane, homes may be damaged or destroyed by high winds and high waves. Debris can break windows and doors, allowing high winds inside the home. In extreme storms, such as Hurricane Andrew, the force of the wind alone can cause weak places in your home to fail.

After Hurricane Andrew, a team of experts examined homes that had failed and ones that had survived. They found four areas that should be checked for weakness—the roof, windows, doors, and if you have one, garage door. We discuss some things you can do to help make your home stronger before the next hurricane strikes. You may need to make some improvements or install temporary wind protection. It is important that you do these projects now, before a hurricane threatens.

While these projects, if done correctly, can make your home safer during a hurricane, they are no guarantee that your home won't be damaged or even destroyed. If you are told by authorities to evacuate, do so immediately, even if you have taken these precautions.

The Roof

During a windstorm, the force of the wind pushes against the outside of your home. That force is passed along from your roof to the exterior walls and finally to the foundation. Homes can be damaged or destroyed when the energy from the wind is not properly transferred to the ground.

The first thing you should do is determine what type of roof you have. Homes with gabled roofs are more likely to suffer damage during a hurricane. A gabled roof looks like an A on the ends, with the outside wall going to the top of the roof. The end wall of a home with gabled roof takes a beating during a hurricane, and those that are not properly braced can collapse, causing major damage to the roof.

In most homes, gabled roofs are built using manufactured trusses. Sheets of roof sheathing, often plywood, are fastened to the trusses with nails or staples, and roofing material is fastened to the sheathing. In many cases, the only thing holding the trusses in place is the plywood on top. This may not be enough to hold the roof in place during a hurricane. Installing additional truss bracing makes your roof's truss system much stronger.

To inspect your roof's bracing, go into the attic. While working in your attic, you should wear clothing that covers your skin, work gloves, a hat, eye protection, and a dust mask. If your attic does not have a floor, be careful to walk only on wood joists, or install boards wide enough to walk on as you work. Notice how the plywood is attached to the truss system. If most of the large nails or staples coming through the sheathing have missed the trusses, consider having the sheathing properly installed.

Truss Bracing

In gabled roofs, truss bracing usually consists of 2x4s that run the length of the roof. If you do not have truss bracing, it should be installed. You can do this yourself or hire a professional. Install 2x4s the length of your roof, overlapping the ends of the 2x4s across two trusses. Braces should be installed 18 inches from the ridge, in the center span, and at the base, with 8 to 10 feet between the braces. Use two 3-inch, 14-gauge wood screws or two 16d (16 penny) galvanized common nails at each truss. Because space in attics is generally limited, screws may be easier to install.

Gable End Bracing

Gable end bracing consists of 2x4s placed in an "X" pattern from the top center of the gable to the bottom center brace of the fourth truss, and from the bottom center of the gable to the top center brace of the fourth truss. Use two 3-inch, 14-gauge wood screws or two 16d galvanized common nails to attach the 2x4s to the gable and to each of the four trusses.

Hurricane Straps

There are many types of roof design. Regardless of your type of roof, hurricane straps are designed to help hold your roof to the walls. While you are in the attic, inspect for hurricane straps of galvanized metal. Hurricane straps may be difficult for homeowners to install. You may need to call a professional to retrofit your home with hurricane straps. Check with your local government building officials to see if hurricane straps are required in your area.

Exterior Doors and Windows

The exterior walls, doors, and windows are the protective shell of your home. If your home's protective shell is broken, high winds can enter and put pressure on your roof and walls, causing damage. You can protect your home by strengthening the doors and windows.

Double Entry Doors

Most double doors have an active and an inactive or fixed door. Check to see how the fixed door is secured at the top and bottom. The bolts or pins that secure most doors are not strong enough.

Some door manufacturers provide reinforcing bolt kits made specifically for their doors. Check with your local building supplies retailer to find out what type of bolt system will work for your door. The door bolt materials should cost from $10 to $40, depending on the type and finish. Doors with windows will need additional protection from flying debris. See the section on storm shutters for how to protect windows.

Double-wide garage doors

Double-wide (two-car) garage doors can pose a problem during hurricanes because they are so large that they wobble as the high winds blow and can pull out of their tracks or collapse from wind pressure. If garage doors fail, high winds can enter your home through the garage and blow out doors, windows, walls, and even the roof.

Certain parts of the country have building codes requiring garage doors to withstand high winds. You should check with local government building officials to see if there are code requirements for garage doors in your area. Some garage doors can be strengthened with retrofit kits. Check with your local building supplies retailer to see if a retrofit kit is available for your garage door. You can expect to pay from $70 to $150 to retrofit your garage door.

Many garage doors can be reinforced at their weakest points. Retrofitting your garage doors involves installing horizontal bracing onto each panel. This horizontal bracing can be part of a kit from the garage door manufacturer. You may also need heavier hinges and stronger center supports and end supports for your door.

Check the track on your garage door. With both hands, grab a section of each track and see if it is loose or if it can be twisted. If so, a stronger track should be installed. Make sure that it is anchored to the 2x4s inside the wall with heavy wood bolts or properly attached to masonry with expansion bolts.

After you have retrofitted your door, it may not be balanced. To check, lower the door about halfway and let go. If it goes up or down, the springs will need adjusting. The springs are dangerous and should be adjusted by a professional.

If you are unable to retrofit your door, you can purchase specially reinforced garage doors designed to withstand winds of up to 120 miles per hour. These doors can cost from $400 to $450 (excluding labor) and should be installed by a professional.

Storm shutters

Installing storm shutters over all exposed windows and other glass surfaces is one of the easiest and most effective ways to protect your home. You should cover all windows, French doors, sliding glass doors, and skylights. There are many types of manufactured storm shutters, check with your local building supplies retailer. If you install manufactured shutters, follow the manufacturer's instructions carefully.

Before installing shutters, check with your local building official to find out if a building permit is required. It is important that you have your shutters ready now, and that you mark and store them so they can be easily installed during a hurricane watch.

Plywood shutters that you make yourself, if installed properly, can offer a high level of protection from flying debris during a hurricane. Plywood shutters can be installed on all types of homes.

Measure each window and each door that has glass, and add 8 inches to both the height and width to provide a 4-inch overlap on each side of the window or door. Sheets of plywood are generally 4x8 feet. Tell your local building supply retailer the size and number of openings you need to cover to determine how many sheets to buy.

To install plywood shutters you will need bolts, wood or masonry anchors, large washers, and 5/8-inch exterior-grade plywood. For windows 3 feet by 4 feet or smaller installed on a wood frame house, use 1/4-inch lag bolts and plastic-coated permanent anchors. The lag bolts should penetrate the wall and frame surrounding the window at least 1 3/4 inches. For larger windows, use 3/8-inch lag bolts that penetrate the wall and frame surrounding the window at least 2 1/2 inches. For windows 3 feet by 4 feet or smaller installed on a masonry house, us 1/4-inch expansion bolts and galvanized permanent expansion anchors. The expansion bolt should penetrate the wall at least 1 1/2 inches. For larger windows, use 3/8-inch expansion bolts that penetrate the wall at least 1 1/2 inches. The tools you will need are a circular or hand saw, a drill with the appropriately sized bits, a hammer, and a wrench to fit the bolts. To be safe, use eye protection and work gloves.

Cut the plywood to the measurements for each opening. Drill holes 2 1/2 inches from the outside edge of the plywood at each corner and at 12-inch intervals. Drill four holes in the center area of the plywood to relieve pressure during a hurricane.

Place the plywood over the opening and mark each hole position on the outside wall. Drill holes with the appropriate size and type of bit for the anchors. Install the anchors, the plywood, and the bolts to make sure they fit properly. On wood-frame houses, make sure that the anchors are secured into the solid wood that frames the door or window and not into the siding or trim. Mark each shutter so you will know where it is to be installed and store them and the bolts in an accessible place.

If the opening is larger than one sheet of plywood, you will need to make shutters with 2x4s at the middle and bottom of the two sheets of plywood, evenly spaced, with the 2-inch side attached to the inside of the storm shutter. Attach the 2x4s to the outside of the storm shutter with 2-inch, 10-gauge wood screws before installing the shutter.

These recommendations are not intended to replace local building code requirements or to serve as the only options for protecting your home from hurricane wind damage. For more information on protecting your home from hurricane wind damage, contact your local building official; your local building supply retailer; or a building professional, such as an engineer, architect, or experienced contractor.

Source: Against the Wind—Protecting Your Home from Hurricane Wind Damage, FEMA, 7 pp. (Also available at http://www.fema.gov.)

Excerpts from the FEMA Document "Surviving the Storm— A Guide to Hurricane Preparedness"

This FEMA document provides advice on how to prepare before and during a hurricane threat. The best preparation is advanced planning before the hurricane season becomes active. Once a hurricane watch is issued, it is frequently too late to protect your home because stores will be running out of supplies. FEMA also provides advice on how to protect pets, mobile homes, and businesses. The importance of acquiring flood insurance through the National Flood Insurance Program is discussed, because most homeowner policies do not cover damage by floods. This federal program provides flood insurance at a reasonable cost to residents whose communities participate in the National Flood Insurance Program. In order to participate, the community must adopt and enforce local floodplain management ordinances designed to reduce the risk of future flood losses.

Know Your Risks

One of the most dramatic, damaging and potentially deadly weather events that occurs in this country is a hurricane. Fortunately, there are measures that can be taken by individuals and communities to reduce vulnerability to hurricane hazards.

During a hurricane, homes, businesses, public buildings, roads and power lines may be damaged or destroyed by high winds and floodwaters. Debris can break windows and doors. Roads and bridges can be washed away by flash flooding or blocked by debris.

The force of wind alone can cause tremendous devastation, toppling trees and power lines and undermining weak areas of buildings.

These storms cost our nation millions, if not billions, of dollars in damage annually. But there are ways to offset such destruction. Simple construction measures, such as placing storm shutters over exposed glass or installing hurricane straps on roofs, have proven effective in lessening damage when hurricanes strike.

Communities can reduce vulnerability to hurricanes by adopting and enforcing building codes for wind and flood resistance. Sound land-use planning also can ensure that structures are not built in high-hazard areas.

Building disaster-resistant communities is an achievable goal. It requires action by individuals, businesses and local governments. Working together, we can reduce the number of lives, property and businesses lost the next time a hurricane strikes.

Prepare a Family Disaster Plan

A well-thought-out plan of action for you and your family can go a long way to reduce potential suffering from any type of disaster that could strike.

Household emergency plans should be kept simple. The best emergency plans are those that are easy to remember.

Be familiar with escape routes. It may be necessary to evacuate your neighborhood. Plan several escape routes for different contingencies.

Maintaining a link to the outside can be crucial. Keep a battery-operated radio and extra batteries on hand. Make sure family members know where the radio is kept.

Post emergency phone numbers (fire, police, ambulance) by the telephone.

Teach children how to call 911 for help.

Know how to turn off utilities.

Identify family meeting places in case you are separated. Choose a place in a building or park outside your neighborhood. Everyone should be clear about this location.

Develop an emergency communication plan. Ask an out-of-state relative or friend to serve as the family's contact. Make sure everyone knows the telephone number of this contact.

Preparedness: Act Before a Hurricane Strikes

Over the past several years, the hurricane warning system has provided time for people to move inland when hurricanes threaten. However, it is becoming more difficult to evacuate people from densely populated areas. Roads are easily overcrowded, particularly during summer tourist season. The problem is compounded by the complacency of people who do not understand the awesome power of the storm. Complacency and delayed action could result in needless loss of life and damage to property.

Before a Hurricane Strikes

Plan a safe evacuation route that will take you 20–50 miles inland. Contact your local emergency management office or Red Cross chapter and ask for the community preparedness plan.

Have disaster supplies on hand, such as:

- Flashlight and extra batteries
- Portable, battery-operated radio and extra batteries
- First-aid kit
- Emergency food and water
- Nonelectric can opener
- Essential medicines

- Cash and credit cards
- Sturdy shoes and a change of clothing

Make sure your family goes over the family disaster plan. (See "Prepare a Family Disaster Plan.")

Make plans for protecting your house, especially the roof, windows and doors. (See accompanying document titled "Protecting Your Home from Hurricane Wind Damage.")

Trim dead or weak branches from trees.

Check into flood insurance. Homeowner policies do not cover damage from flooding that often accompanies hurricanes. Call your local insurance agent for information on the National Flood Insurance Program.

When a Hurricane Watch or Warning Is Issued

Listen to radio or television for hurricane progress reports.

Check your emergency supplies. Store drinking water in clean bathtubs, jugs, bottles and cooking utensils.

Bring in outdoor objects such as lawn furniture, toys and garden tools; anchor objects that cannot be brought inside but that could be wind-tossed. Remove outdoor antennas, if possible.

Secure your home by closing or installing window shutters.

Turn the refrigerator and freezer to the coldest setting. Open only when necessary and close quickly.

Fuel your car. If you have a boat, moor it securely or move it to a designated safe place.

Store valuables and personal papers in a waterproof container.

If Evacuation Is Necessary

If officials order evacuation, leave as soon as possible. Avoid flooded roads and watch for washed-out bridges.

Secure your home. Unplug appliances and turn off electricity and the main water valve. If time permits, elevate furniture to protect it from flooding or move it to a higher floor.

Take your pre-assembled emergency supplies and warm, protective clothing.

After the Hurricane Passes

Return home only after authorities say it is safe to do so.

Beware of downed or loose power lines. Report them immediately to the power-company, police or fire department.

Enter your home with caution. Open windows and doors to ventilate or dry your home.

Check for gas leaks. If you smell gas or hear a blowing or hissing noise, quickly leave the building and leave the doors open. Call the gas-company.

Look for electrical system damage. If you see sparks or frayed wires, turn off electricity at the main fuse box. If you have to step in water to reach the electric box, call an electrician for advice.

Check for sewage and water-line damage. If you suspect there is such damage, call the water company and avoid using water or toilets until they come.

Take pictures of the damage for insurance claims.

Mobile Homes Require Special Precautions

Mobile homes are particularly vulnerable to hurricane-force winds. Anchor the mobile home with over-the-top or frame ties. When a storm threatens, do what you can to secure your home, and then take refuge with friends or relatives or at a public shelter.

Before you leave, take the following precautions:

- Pack breakables in boxes and put them on the floor.
- Remove mirrors and tape them. Wrap mirrors and lamps in blankets and place them in the bathtub or shower.
- Install shutters or precut plywood on all windows.
- Disconnect electricity, sewer and water lines. Shut off propane tanks and leave them outside after anchoring them securely.
- Store awnings, folding furniture, trashcans and other such loose outdoor objects.

Make Plans for Your Pets

In planning for the hurricane season, do not forget your pets. If you evacuate your home, do not leave pets behind.

The Humane Society of the United States urges pet owners to make arrangements to evacuate their animals.

Be sure you have up-to-date identification tags, a pet carrier and a leash for them.

Most emergency shelters will not accept pets. In the event of evacuation, make alternative arrangements for pets, such as with family friends, veterinarians or kennels in safe locations.

Send medicine, food, feeding information and other supplies with them.

Planning Could Save Your Business

If a hurricane is threatening the area where your business is located, you can take actions ahead of time that will save damage and lost productivity.

Clear out areas with extensive glass frontage as much as possible. If you have shutters, use them; otherwise, use precut plywood to board up doors and windows. Remove outdoor hanging signs.

Bring inside or secure any objects that might become airborne and cause damage in strong winds.

Secure showcases. Use plywood to protect glass showcases or if possible, turn the glass side toward an inside wall.

Store as much merchandise as high as possible off the floor, especially goods that could be in short supply after the storm.

Move merchandise that cannot be stored away from glass and cover it with tarpaulins or heavy plastic.

Secure generators, along with fuel needed for their operation.

Secure all goods in warehouses above the water level, and place sandbags in spaces where water could enter.

Remove papers from lower drawers of desks and file cabinets and place them in plastic bags on top of the cabinets.

Turn off water heaters, stoves, pilot lights and other burners.

Danger: Flash Floods

Nearly half of all flash flood fatalities are automobile related.

Water weighs 62.4 lbs. per cubic foot and typically flows downstream at 6 to 12 mph.

When a vehicle stalls in water, the water's momentum is transferred to the car. For each foot the water rises, 500 lbs. of lateral force are applied to the car.

But the biggest factor is buoyancy. For each foot the water rises up the side of the car, the car displaces 1,500 lbs. of water. In effect, the car weighs 1,500 lbs. less for each foot the water rises.

Two feet of water will carry away most automobiles.

Source: Surviving the Storms—A Guide to Hurricane Preparedness. FEMA, 6 pp. Washington, DC. (Available from http://www.fema.gov.)

Excerpts from the FEMA Document "Recovering Flood Damaged Property"

The hurricane storm surge in exposed coastlines, as well as flooding further inland, are common occurrences. This article discusses safety precautions when recovering from a flood.

Safety Tips for Flood Victims

Do not use electrical appliances that have been wet. Water can damage the motors in electrical appliances, such as furnaces, freezers, refrigerators, washing machines, and dryers.

If electrical appliances have been under water, have them dried out and reconditioned by a qualified service repairman. Do not turn on damaged electrical appliances because the electrical parts can become grounded and pose an electric shock hazard or overheat and cause a fire. Before flipping a switch or plugging in an appliance, have an electrician check the house wiring and appliance to make sure it is safe to use.

Electricity and water don't mix. Use a ground fault circuit interrupter (GFCI) to help prevent electrocutions and electrical shock injuries. Portable GFCIs require no tools to install and are available at prices ranging from $12 to $30.

When using a "wet-dry vacuum cleaner," be sure to follow the manufacturer's instructions to avoid electric shock. Do not allow the power cord connections to become wet. Do not remove or bypass the ground pin on the three-prong plug. Use a GFCI to prevent electrocution.

NEVER remove or bypass the ground pin on a three-pronged plug in order to insert it into a non-grounded outlet.

NEVER allow the connection between the machine's power cord and the extension cord to lie in water.

To prevent a gas explosion and fire, have gas appliances (natural gas and LP gas) inspected and cleaned after flooding.

If gas appliances have been under water, have them inspected and cleaned and their gas controls replaced. The gas company or a qualified appliance repair person or plumber should do this work. Water can damage gas controls so that safety features are blocked, even if the gas controls appear to operate properly. If you suspect a gas leak, don't light a match, use any electrical appliance, turn lights on or off, or use the phone. These may produce sparks. Sniff for gas leaks, starting at the water heater. If you smell gas or hear gas escaping, turn off the main valve, open windows, leave the area immediately, and call the gas company or a qualified appliance repair person or plumber for repairs.

Never store flammable materials near any gas appliance or equipment.

Check to make sure your smoke detector is functioning. Smoke detectors can save your life in a fire. Check the battery frequently to make sure it is operating. Fire extinguishers also are a good idea.

Gasoline is made to explode!

Never use gasoline around ignition sources such as cigarettes, matches, lighters, water heaters, or electric sparks. Gasoline vapors can travel and be ignited by pilot light or other ignition sources. Make sure that gasoline powered generators are away from easily combustible materials.

Chain saws can cause serious injuries. Chain saws can be hazardous, especially if they "kick back." To help reduce this hazard, make sure that your chain saw is equipped with the low-kickback chain. Look for other safety features on chain saws, including hand guard, safety tip, chain brake, vibration reduction system, spark arrestor on gasoline models, trigger or throttle lockout, chain catcher, and bumper spikes. Always wear shoes, gloves, and protective glasses. On new saws, look for certification to the ANSI B-175.1 standard.

When cleaning up from a flood, store medicines and chemicals away from young children. Poisonings can happen when young children swallow medicines and household chemicals.

Keep household chemicals and medicines locked up away from children. Use the child resistant closures that come on most medicines and chemicals.

Burning charcoal gives off carbon monoxide. Carbon monoxide has no odor and can kill you. Never burn charcoal inside homes, tents, campers, vans, cars, trucks, garages, or mobile homes.

WARNING: Submerged gas control valves, circuit breakers, and fuses pose explosion and fire hazard!

Replace all gas control valves, circuit breakers, and fuses that have been under water:

GAS CONTROL VALVES on furnaces, water heaters, and other gas appliances that have been under water are unfit for continued use. If they are used, they could cause a fire or an explosion. Silt and corrosion from flood water can damage internal components of control valves and prevent proper operation. Gas can leak and result in an explosion or fire. Replace ALL gas control valves that have been under water.

ELECTRIC CIRCUIT BREAKERS AND FUSES can malfunction when water and silt get inside. Discard ALL circuit breakers and fuses that have been submerged.

Source: Recovering Flood Damaged Property, FEMA. Washington, DC, 1998. (Available from http://www.fema.gov.)

Directory of Organizations

This chapter describes some organizations relevant to hurricane research and forecasting. Other organizations are included because of their importance to meteorology as a whole or because of their commitment to disaster mitigation or assistance. A description of each group's function and history is included, as well as its address, point of contact, and web page. Relevant publications are also listed.

American Meteorological Society
45 Beacon Street
Boston, MA 02108-3693
(617) 227-2426
http://www.ametsoc.org/AMS/

The American Meteorological Society (AMS) is a nonprofit, professional society whose objectives are the development and dissemination of knowledge of the atmospheric and related oceanic and hydrologic sciences and the advancement of its professional applications. Charles Franklin Brooks of the Blue Hill Observatory in Milton, Massachusetts, founded the AMS in 1919. Its initial membership came primarily from the U.S. Signal Corps and U.S. Weather Bureau. Its initial publication, the *Bulletin of the American Mete-*

orological Society, was meant to serve as a supplement to the *Monthly Weather Review,* which, at the time, was published by the U.S. Weather Bureau.

During the 1930s and 1940s, however, because of the key role meteorologists played in support of military activities and because of associated technology developments, the AMS grew substantially in numbers and in purpose. Carl-Gustav Rossby served as president of the society for 1944 and 1945 and developed the framework for the society's first scientific journal, the *Journal of Meteorology,* which later split into the two current AMS journals: the *Journal of Applied Meteorology* and the *Journal of the Atmospheric Sciences.*

This role as a scientific and professional organization serving the atmospheric and related sciences continues today. The AMS now publishes eight scientific journals and an abstract journal, in addition to the *Bulletin,* and sponsors and organizes over a dozen scientific conferences each year. It has published almost 50 monographs in its continuing series, as well as many other books and educational materials. Most of these journals are directed toward an audience with a bachelor of science degree in meteorology, and many are written for those with advanced degrees. However, the *Bulletin* sometimes contains articles written for the general public, and many of the educational materials are quite readable. The AMS administers two professional certification programs, the Radio and Television Seal of Approval and the Certified Consulting Meteorologist programs, and also offers an array of undergraduate scholarships and graduate fellowships to support students pursuing careers in the atmospheric and related oceanic and hydrologic sciences. Local AMS chapters for many cities and universities also exist, providing a means for community interaction between those interested in meteorology and professional meteorologists.

One may join the AMS under the classification of "member" if he or she is a professional in the field of atmospheric or related oceanic or hydrologic sciences holding a baccalaureate or higher degree. If one does not meet this criteria but is interested in joining the AMS, he or she may apply as an "associate member." Other classifications for students also exist. As of this writing, annual dues were $30, with many optional purchases, as detailed on their website.

Publications: For professional meteorologists, the following journals are published by the AMS: the *Bulletin of the American Me-*

teorological Society, Weather and Forecasting, Journal of Climate, Monthly Weather Review, Journal of Physical Oceanography, Journal of the Atmospheric Sciences, Journal of Applied Meteorology, Journal of Atmospheric and Oceanic Technology, Journal of Hydrometeorology, Earth Interactions, and *Meteorological and Geastrophysical Abstracts.*

To promote studies in the atmospheric sciences at elementary and secondary school levels, AMS has created Project ATMOSPHERE, an educational program designed to foster the teaching of atmospheric topics across the grades K-12. Its main goal is to promote interest and literacy in science, technology, and mathematics at the precollege level. A similar educational program also exists for oceanography, called the Maury Project. Both projects provide a plethora of educational-related publications, with classroom exercises, on a variety of weather and ocean topics, including hurricanes.

The AMS also publishes many monographs on weather topics. The following partially or exclusively discuss hurricanes: *Historical Essays on Meteorology; Early American Hurricanes: 1492–1870; Tropical Cyclones—Their Evolution, Structure, and Effects; Glossary of Weather and Climate;* and *The 1938 Hurricane.*

American Red Cross
P.O. Box 37243
Washington, DC 20013
Phone: 1-800-435-7669
http://www.redcross.org

The American Red Cross is a humanitarian organization, led by volunteers, that provides relief to victims of disasters and helps prevent, prepare for, and respond to emergencies. It is the largest volunteer emergency service organization in the United States, with more than 1,300 chapters nationwide, 38 blood services regions, 18 tissue services centers, plus hundreds of stations on U.S. military installations around the world. In 1998 the organization had 1.33 million volunteers and less than 30,000 paid staff. The Red Cross is not a U.S. government agency but a private, non-profit organization relying on charitable donations for funding. The American Red Cross is part of the larger International Red Cross, which comprises different Red Cross organizations in various countries.

The Red Cross was founded by Henri Dunant in June 1859 when he visited the battlefields of Solferino in Italy and found thousands of wounded and dying with no medical care. He organized a group to assist and wrote a book about the experience afterwards. This book motivated a movement to authorize medical personnel to assist wounded soldiers and to use an emblem (a red cross on a white armlet) to protect them. A second component of the movement was to have governments sign treaties to protect these workers and to have national volunteer societies formed to carry out Dunant's mission. During this movement, a volunteer named Clara Barton cared for soldiers during the American Civil War. After the Civil War, Barton visited Europe and learned of Dunant's movement. She returned home deeply committed to persuading the United States to participate in the movement. In 1864, the United States and many other countries signed such a treaty, known as the Geneva Convention. It established that "wounded or sick combatants, to whatever nation they belong, shall be collected and cared for" by personnel wearing the Red Cross emblem and that persons and facilities bearing the symbol are protected from attack. On May 21, 1881, Barton founded the American Red Cross, and in 1900 the U.S. Congress chartered the organization to provide services to the U.S. military forces and to victims at home and abroad. In a typical year, the Red Cross responds to more than 66,000 natural and man-made disasters, including hurricanes, floods, earthquakes, tornadoes, fires, hazardous materials spills, civil disturbances, explosions, and transportation accidents.

Publications: The Red Cross publishes several pamphlets related to hurricane preparedness and hurricane relief, including: *Disaster Preparedness for People with Disabilities, Partners in Action Newsletter,* and a photo feature entitled *Hurricanes: The Faces of Disaster.* Many of these pamphlets can be viewed (or downloaded as pdf files) from their website. Other pamphlets can be purchased for a nominal fee, including *Jason and Robin's Awesome Hurricane Adventure* color workbook, *Are You Ready for A Hurricane?,* and *Hurricanes . . . Unleashing Nature's Fury!* These last two publications are also available from the National Weather Service. The American Red Cross also sells videos on hurricane preparedness.

Cooperative Institute for Meteorological Satellite Studies (CIMSS) at the University of Wisconsin-Madison
1225 West Daytona Street
Madison, WI 53706

Phone: (608) 263-7435
Fax: (608) 262-5974
http://cimss.ssec.wisc.edu

The Cooperative Institute for Meteorological Satellite Studies (CIMSS) was established in 1980 to formalize and support cooperative research between the National Oceanic and Atmospheric Administration's (NOAA) National Environmental Satellite, Data, and Information Service (NESDIS) and the University of Wisconsin-Madison's Space Science and Engineering Center. Sponsorship and membership of CIMSS was expanded to include the National Aeronautics and Space Administration (NASA) in 1989. During the 1980s, a need emerged for joint federal-university research centers to support the NOAA weather research program. CIMSS was established to focus on the development and testing of the operational utility of new weather satellite observing systems to improve weather forecasts.

CIMSS develops techniques for using geostationary weather satellite observations to improve forecasts of severe storms, including tornadoes and hurricanes. In particular, CIMSS has developed a methodology for computing wind speed and direction by tracking individual cloud elements from satellite imagery. This procedure produces wind information in data-sparse regions such as the tropical oceans. These winds are used to identify steering currents and to visualize atmospheric features that affect tropical cyclone genesis and intensification. Researchers have begun to assimilate these data into computer models in the hope it will improve weather forecasts, including hurricane track predictions.

CIMSS also participates in satellite instrument design, international field programs, and computer visualization of satellite measurements. For example, software called the Man computer Interactive Data Access System (McIDAS) displays, enhances, and animates satellite data and can also lay meteorological measurements over the imagery. McIDAS is still actively used today by many governmental and university institutions, including NHC.

Media representatives desiring additional information should contact CIMSS's public information specialist at (608) 263-3373.

Publications: The institute's web page is the best source for information about hurricane research at CIMSS. On the web page under the "Tropical Cyclones" link, worldwide satellite images of hurricanes are displayed in a variety of formats, as are the winds. Infrared, visible, and water vapor satellite images are also available. Hurricane forecasters from around the world use this state-of-the-art website.

Federal Emergency Management Agency (FEMA)
Federal Center Plaza
500 C. Street S.W.
Washington, DC 20472
http://www.fema.gov

The Federal Emergency Management Agency (FEMA) is an independent agency of the federal government, reporting to the president. Since its founding in 1979, FEMA's mission has been to reduce loss of life and property and protect our nation's critical infrastructure from all types of hazards (including hurricanes) through a comprehensive, risk-based, emergency management program of mitigation, preparedness, response, and recovery. FEMA has about 2,600 full-time employees and nearly 4,000 standby disaster assistance employees who are available to help after disasters. Often FEMA works in partnership with other organizations that are part of the nation's emergency management system. These partners include state and local emergency management agencies, 27 federal agencies, and the American Red Cross.

FEMA is called in to help when the president declares a disaster. Disasters are "declared" after hurricanes, tornadoes, floods, earthquakes, or other similar events strike a community. The governor of the state must ask for help from the president before FEMA can respond. FEMA helps disaster victims find a place to stay if their homes were damaged or destroyed. FEMA also helps repair homes and works with city officials to fix public buildings that have been damaged.

FEMA is also involved in disaster mitigation by teaching people how to prepare for a disaster and how to make their homes as safe as possible. FEMA works with communities to help them build safer, stronger buildings that are less likely to be damaged. FEMA also trains firefighters and emergency workers and runs a flood insurance program.

FEMA also has 10 regional offices, and 2 area offices. Each region serves several states, and regional staff work directly with the states to help plan for disasters, develop mitigation programs, and meet needs when major disasters occur. These regions are shown on FEMA's website. For more immediate responses to emergency management issues, it may be better to contact the nearest regional office.

Publications: FEMA's web page contains extensive documentation on hurricane preparedness and hurricane recovery issues. Many publications are also available, including *Hurricane— Floods: Safety Tips for Coastal and Inland Flooding, Safety Tips for Hurricanes, Hurricane Awareness—Action Guidelines for Senior Citizens, Hurricane Awareness—Action Guidelines for School Children, Against the Wind—Protecting Your Home from Hurricane Wind Damage,* and *After a Flood—The First Steps.*

Geophysical Fluid Dynamics Laboratory (GFDL)
P.O. Box 308
Princeton, NJ 08542-0308
(609) 452-6500
http://www.gfdl.gov

The Geophysical Fluid Dynamics Laboratory (GFDL) is another NOAA research laboratory, and its goal is to understand and predict the earth's climate and weather, including the impact of human activities. GFDL conducts leading-edge research on many topics of great practical value, including weather and hurricane forecasts, El Niño prediction, stratospheric ozone depletion, and global warming.

With respect to hurricanes, GFDL's mission is (1) to understand the genesis, development, and decay of tropical disturbances by investigating the thermo-hydrodynamical processes using numerical simulation models; (2) to study small-scale features of hurricane systems, such as the collective role of deep convection, the exchange of physical quantities at the lower boundary, and the formation of organized spiral bands; and (3) to investigate the capability of numerical models in predicting hurricane movement and intensity and to facilitate their conversion to operational use.

GFDL's hurricane model provides forecast track and intensity guidance to the National Hurricane Center. A similar version is used by the Navy for the western Pacific Ocean and is called the GFDN model. GFDL's model, on average, produces the smallest track errors compared to other operational models.

Publications: GFDL's web page is the best source for information about hurricane research at GFDL. In addition, one section of the website contains online bibliographies of researchers' abstracts going back as far as 1965.

Hurricane Hunters
403rd Wing Public Affairs
701 Fisher Street, Suite 103
Keesler AFB, MS 39534-2572
Phone: (228) 377-2056
Fax (228) 377-0755
http://www.hurricanehunters.com

The 53rd Weather Reconnaissance Squadron, known as the Hurricane Hunters of the Air Force Reserve, is the only Department of Defense organization flying into tropical storms and hurricanes on a routine basis. The 53rd Weather Reconnaissance Squadron, a component of the 403rd Wing located at Keesler Air Force Base in Biloxi, Mississippi, is the only unit in the world flying into hurricanes on a routine basis.

The mission of the Hurricane Hunters is to recruit, organize, and train assigned personnel to perform aerial weather reconnaissance. During the hurricane season from June 1 to November 30, they provide surveillance of tropical disturbances, tropical depressions, tropical storms, and hurricanes in the Atlantic (west of 55 deg W), Caribbean, and Gulf of Mexico for the National Hurricane Center in Miami, Florida. They also may fly storms for the Central Pacific Hurricane Center in Honolulu, Hawaii.

From November 1 through April 15, the unit also flies winter storms off both coasts of the United States in support of the National Center for Environmental Prediction. These missions are flown at high altitude (30,000 feet) and can be just as challenging as the hurricane missions, with turbulence, lightning, and icing. Weather data are needed from areas where it is impractical or impossible to operate ground observation stations or where weather satellites cannot provide complete information. To fill this need, Air Force aerial weather reconnaissance aircraft fly long routes over the ocean to collect and transmit weather observations and atmospheric soundings.

As of the late 1990s, a typical hurricane warning cost an estimated $192 million, considering the costs of preparation, evacuation, and lost commerce. Narrowing the warning area could save $300,000 per mile or more and could also lend greater credibility to forecasts and foster more controlled and limited coastal evacuations. Furthermore, as coastal populations continue to grow, evacuation decisions need to be made earlier. Studies have shown the high accuracy data from our Air Force Reserve and NOAA aircraft have improved the forecast accuracy by about 25 percent. Air crews in these storms also have detected sudden, dangerous changes in hurricane intensity and movement, which are currently very difficult to detect by satellite alone.

The reservists bring a wide variety of experience with them; their primary occupations include airline pilot, meteorologist, teacher, electrician, computer systems operator, law enforcement officer, firefighter, medical doctor, veterinarian, and businessman. Although most live in the Gulf Coast area, the reservists are drawn from 15 states and frequently travel at their own expense to perform their duties.

Members of the media may get permission to fly on one of the reconnaissance flights into a hurricane. Please note that only media members are allowed to accompany the Hurricane Hunters, and all other requests will be turned down.

Publications: The Hurricane Hunters' website is the best source of information about this organization. In addition, they publish a pamphlet titled *53rd Weather Reconnaissance Squadron—Hurricane Hunters.*

Hurricane Research Division (HRD)

NOAA/AOML/HRD
4301 Rickenbacker Causeway
Miami, FL 33149
Phone: (305) 361-4400
Fax: (305) 361-4402
http://www.aoml.noaa.gov/hrd

The Hurricane Research Division (HRD) is a part of the Atlantic Oceanographic and Meteorological Laboratory (AOML) within the National Oceanic and Atmospheric Administration (NOAA) of the U.S. Department of Commerce. HRD is NOAA's primary focus for studies of tropical cyclones and tropical thunderstorms. HRD's mission is:

1. To improve forecasting through physical understanding of the structure and dynamics of hurricanes.
2. To investigate specific research topics as specified annually by the Hurricane Field Program. This research is fa-

cilitated using two NOAA WP-3D research aircraft (for analysis of tropical systems between 5,000 and 10,000 feet) and a Gulfstream IV-SP operational jet (for analysis at 20,000 feet and higher).

3. To interact with the National Hurricane Center, the National Center for Environmental Prediction (NCEP), and the Geophysical Fluid Dynamics Laboratory (GFDL) in hurricane modeling research.

4. To participate in international tropical meteorological field programs and national research programs.

The aircraft are provided by the NOAA Aircraft Operations Center (AOC). This facility is discussed later in this chapter.

Publications: AOML produces a monthly newsletter called *AOML Keynotes,* which may also be viewed on their website in pdf format (go to http://www.aoml.noaa.gov, then look in the "AOML Overview" link). The web page contains a popular Frequently Asked Question section on hurricanes and on El Niño. In addition, historical data sets containing positions and intensities for past Atlantic, East Pacific, West Pacific, and Australian tropical cyclones (commonly called besttrack data sets) can be downloaded for free. The Atlantic besttrack data set is quite extensive, going back to the year 1886, and is used for analyzing past hurricane tracks as well as for climate studies. See the section "National Weather Service" for a detailed listing of other related NOAA pamphlets.

Joint Typhoon Warning Center (JTWC)
Naval Pacific Meteorology and Oceanography Center
Box 113
Pearl Harbor, HI 96860-5050
Phone: (808) 474-5301
http://www.npmoc.navy.mil

The Joint Typhoon Warning Center (JTWC) provides hurricane forecasting support to the U.S. military and to allies within the Pacific and Indian Ocean basins under the auspices of the Naval Pacific Meteorology and Oceanography Center in Hawaii. JTWC was founded May 1, 1959, when the U.S. commander-in-chief of the Pacific forces directed that a single hurricane warning center be established for the western North Pacific region. JTWC is a combined Air Force/Navy organization, in which the Air Force provides reconnaissance flight and forecasters and the Navy provides the facility, equipment, and forecasters.

The mission of JTWC includes:

1. Continuous monitoring of all tropical weather activity in the Northern and Southern Hemispheres from 180 degrees E to the east coast of Africa, and the prompt issuance of appropriate advisories and alerts when typhoon development is anticipated
2. Issuance of warnings for all tropical storms and typhoons in this region
3. Post-storm analysis of hurricanes occurring within the western North Pacific and North Indian Oceans
4. Cooperation with the Naval Research Laboratory in Monterey, California, on operational evaluation of hurricane models and forecast aids, and the development of new techniques to support operational forecast requirements

Publications: Every year, the JTWC staff prepares an *Annual Tropical Cyclone Report* summarizing hurricane activity in the region. Annual reports of past typhoon seasons starting in 1996 may also be read on their website. In addition, a historical data set containing positions and intensities for West Pacific and Australian hurricanes (commonly called besttrack data sets) can be downloaded for free. JTWC's homepage also contains current typhoon information.

National Aeronautics and Space Administration (NASA)
Washington, DC 20546-0001
Phone: (202) 358-0000
http://www.hq.nasa.gov

Global Hydrology and Climate Center
977 Explorer Boulevard
Huntsville AL 35806
Phone: (256) 922-5700
Fax: (256) 922-5723
http://wwwghcc.msfc.nasa.gov/

Office of Public Affairs, Code 130
NASA Goddard Space Flight Center

Greenbelt, MD 20771
Phone: (301) 286-8955
Fax: (301) 286-1671
http://www.gsfc.nasa.gov

TRMM Project Scientist, Code 912
NASA Goddard Space Flight Center
Greenbelt, MD 20771
Phone: (301) 614-6354
Fax: (301) 614-5492
http://trmm.gsfc.nasa.gov

Mesoscale Atmospheric Processes Branch, Code 912
Laboratory for Atmospheres
Earth Science Directorate
NASA Goddard Space Flight Center
Greenbelt, MD 20771
Phone: (301) 614-6296
Fax: (301) 614-5492
http://rsd.gsfc.nasa.gov/912/code912

The National Aeronautics and Space Administration (NASA) was established by the National Aeronautics and Space Act of 1958 by Congress "to provide for research into problems of flight within and outside the earth's atmosphere, and for other purposes." While NASA certainly contains a rich space flight history, the organization also conducts research on remote sensing, climate topics, and weather topics (including hurricanes). Different NASA organizations, spread throughout the country, sponsor specialized weather research programs. With regard to recent hurricane research, the Third Convection and Moisture Experiment (CAMEX3) was devoted to the study of hurricane tracking and intensification using NASA-funded aircraft remote sensing instrumentation. CAMEX3 was conducted under the auspices of the Global Hydrology and Climate Center, a cooperative venture of NASA and the University of Alabama in Huntsville. Another hurricane-related program is the Tropical Rainfall Measuring Mission (TRMM) at Goddard Space Flight Center, which, in addition to measuring tropical rainfall using a suite of remote sensing instruments on a satellite, can estimate the vertical cloud structure in hurricanes. Another unit at Goddard, called the Mesoscale Atmospheric Processes Branch, conducts research on the understanding of atmospheric processes using satellite, aircraft, surface-based remote sensing instruments, and computer simulations, with a particular focus on smaller-scale weather phenomena, such as clouds, cyclones, and hurricanes.

Publications: There is no shortage of publications and brochures at NASA, and all NASA websites contain a wealth of information. For information about CAMEX3, the Global Hydrology and Climate Center publishes a brochure entitled *Scientist's Notebook: Seeking New Insight on Hurricanes.* For information about TRMM, Goddard Space Flight Center, under the auspices of the Mission to Planet Earth office, publishes *Tropical Rainfall Measuring Mission Office.* Another useful information packet is the *Mesoscale Atmospheric Processes Branch* brochure. Each organization's website also contains additional information. Teachers and students will be interested in *NASA's Earth Science Enterprise 1998 Education Catalog.* This (and future annual catalogs) describes all of NASA's many educational outreach programs from the Earth Science Enterprise office (contact NASA headquarters for more information, or see http://www.earth.nasa.gov).

National Centers for Environmental Prediction (NCEP)
5200 Auth Road
Washington, DC 20233
http://www.ncep.noaa.gov

The National Centers for Environmental Prediction (NCEP) was established in 1958 as the National Meteorological Center. Since the center's beginning, operational weather forecasting has transformed from an infant discipline into a mature science. Under the auspices of the National Oceanic and Atmospheric Administration's National Weather Service, NCEP provides operational forecast products worldwide and is the starting point for nearly all weather forecasts in the United States. It comprises nine national forecast centers. Each center has a specific responsibility for a portion of the NCEP products and services suite. Seven of the centers provide direct products to users, and two of the centers provide essential support by developing and running complex computer models of the atmosphere on the world's fastest computers, called supercomputers. One of the centers providing direct forecast products is the Tropical Prediction Center, which contains the National Hurricane Center.

Phone calls may be directed to the NOAA Office of Public and Constituent Affairs at (202) 482-6090 (see NOAA section for more details) or the National Weather Service (see National Weather Service section for details).

Publications: A brochure titled *National Centers for Environmental Prediction* is available that describes the individual functions of the nine centers. See the section "National Weather Service" for a detailed listing of other related NOAA pamphlets. The NCEP website is also an excellent source for information.

National Climatic Data Center (NCDC)
Federal Building
151 Patton Avenue
Asheville, NC 28801-5001
Phone: (828) 271-4800
Fax: (828) 271-4876
http://www.ncdc.noaa.gov

The National Climatic Data Center (NCDC) is one of several agencies in the National Environmental Satellite, Data, and Information Service (NESDIS). NESDIS administers the development and use of all operational civilian satellite-based environmental remote sensing systems and the national and international acquisition, processing, dissemination, and exchange of environmental data. NCDC is one of the more popular NESDIS agencies because it contains the world's largest archive of weather data and weather records. NCDC produces several publications and photos of hurricanes, as well as data related to hurricane measurements.

Publications: Many data sets and products are available from their website. A valuable and popular book, *Tropical Cyclones of the North Atlantic Ocean, 1871–1992,* provides historical track information and statistical summaries. For information about a specific hurricane, the publication *Storm Data* is highly recommended. A publication titled *Products and Services Guide* contains detailed information about all of NCDC's products, data sets, and publications.

National Oceanographic and Atmospheric Administration (NOAA)
U.S. Department of Commerce
14th Street & Constitution Ave., NW
Room 6013

Washington, DC 20230
Phone: (202) 482-6090
Fax: (202) 482-3154
http://www.noaa.gov

The mission of the National Oceanographic and Atmospheric Administration (NOAA) is to describe and predict changes in the earth's environment and to conserve and manage wisely U.S. coastal and marine resources. NOAA is subdivided into the following five branches: NOAA Fisheries; the National Environmental Satellite, Data, and Information Service (NESDIS); the National Weather Service (NWS); the Office of Oceanic and Atmospheric Research; and the National Ocean Service. The NWS is actively involved in hurricane forecasting, NESDIS provides satellite information and archives weather data, and the Office of Oceanic and Atmospheric Research contains several laboratories where weather research is conducted, including the Hurricane Research Division and the Geophysical Fluid Dynamics Laboratory. Many of the organizations are described elsewhere in this chapter.

NOAA also participates in the Global Learning and Observations to Benefit the Environment (GLOBE) program, which is a worldwide network of students, teachers, and scientists working together to study and understand the environment. GLOBE students make environmental observations at or near their schools and report their data through the Internet to the scientists, who provide feedback to the students about their data.

Publications: NOAA publishes a monthly newsletter titled *NOAANews*, which may also be read on their website at http://www.noaanews.noaa.gov. NOAA Public Affairs also publishes a series of brochures known as the *NOAA Backgrounder*, which contain detailed information about specific NOAA organizations. Many publications are available from NOAA, including one booklet titled *NOAA-K*, which describes their current satellite technology, and *Hazards Support Activities*, which describes their satellite and archived data of weather events, including hurricanes. See the section "National Weather Service" for a detailed listing of other related NOAA pamphlets.

The NOAA homepage provides numerous links related to weather education, as well as links to their branches, which contain more specialized information. As one browses on this website, one finds numerous satellite photos, graphics, and tables.

National Weather Association
6704 Wolke Court
Montgomery, AL 36116-2134
Phone or Fax : (334) 213-0388
Email: NatWeaAsoc@aol.com
http://www.nwas.org

The NWA's mission is to support and promote excellence in operational meteorology and related activities. The NWA started as a nonprofit organization, incorporated in Washington, D.C., in 1975 mainly to serve individuals interested in operational meteorology and related activities. It has grown to over 2,500 members, 40 corporate members, and over 250 subscribers, including many colleges, universities, and weather service agencies. There are no restrictions on membership except age; applicants should be 18 years old or older. Members receive publications as described below, discounts on other NWA publications, and discounted registration fees for the NWA Annual Meeting. The NWA has an annual awards program and offers the NWA Seal of Approval to qualifying radio and television weathercasters to promote standards and quality broadcasting.

To join the NWA, print out the application form available from their website and send it to the NWA office, or request an application from the NWA office. One can also join by simply sending into the NWA office a short letter of request with your full name and mailing address, and other information that is helpful to the NWA: name of employer or name of school; phone numbers for home, office, and fax; and e-mail address. A check or money order in U.S. funds is also required to cover the annual dues. As of this writing, first year dues are $25 ($12.50 for full-time students). Follow-on year dues are $28 ($14 for full-time students).

Publications: Members receive the quarterly journal *National Weather Digest* and a monthly newsletter. Nonmembers may also purchase these publications. Although these NWA publications are not as technical as the AMS publications, in general they are written for operational meteorologists who have had at least a few meteorology classes and some forecasting experience, and only those with a meteorology degree will really appreciate the articles. However, many of the books available from NWA are nontechnical and suitable for all people interested in a variety of weather topics. The NWA also provides grants to K-12 teachers to help improve the education of their students in meteorology.

National Weather Service (NWS)
1325 East-West Highway
Silver Spring, MD 20910
http://www.nws.noaa.gov
Southern Region Headquarters: (817) 978-4613 ext. 140
Western Region Headquarters: (801) 524-5692
Eastern Region Headquarters: (516) 244-0166
Central Region Headquarters: (816) 426-7621
One may also call the nearest NWS office for local information.

The National Weather Service's mission is to provide weather, hydrologic, and climate forecasts and warnings for the United States, its territories, adjacent waters, and ocean areas, for the protection of life and property and the enhancement of the national economy. Television weathercasters and private meteorology companies prepare their forecasts using the basic forecast and weather observation data the NWS issues continuously all day. In addition, NWS data and products are used by private meteorologists for the provision of specialized services, including consultant work and environmental impact studies.

Another interesting component of the NWS is the Techniques Development Laboratory, which conducts applied research in techniques for improving weather forecasts and product generation. This branch works on the development of storm surge forecasts and the development of short-term forecasts of severe weather events, including hurricanes, using Doppler radar, satellites, and other state-of-the-art platforms. Their website is http://tgsv5.nws.noaa.gov/tdl/.

Publications: Numerous pamphlets and brochures are available at local NWS offices or by calling NOAA or the NWS regional headquarters, including the following: *Survival in a Hurricane; Flash Flood; Flash Floods and Floods . . .The Awesome Power!; River and Flood Program (Hydrologic Services Program); National Centers for Environmental Prediction; Mariner's Guide to Marine Weather Services; Hurricanes . . .Unleashing Nature's Fury; NOAA Weather Radio Pamphlet; NOAA Weather Radio . . .The Voice of the National Weather Service; Atlantic Hurricane Tracking Map; Pacific Hurricane Tracking Map; Atlantic Hurricane Names: 1999–2004; Pacific Hurricane Names: 1999–2004;* and *Hawaiian Hurricane Safety Measures with Central Pacific Tracking Chart.* The following pamphlets are available from both the NWS and the Red Cross: *Hurricane! A Familiarization Booklet* and *Red Cross—Are You Ready for a Flood or Flash Flood?*

Many of these pamphlets can be viewed (or downloaded as pdf files) from the NWS website. This website is an excellent source for current weather information and provides many links to hurricane information.

The Techniques Development Laboratory produces a newsletter every three months titled *Techniques Development Laboratory Quarterly Report,* which may also be read on their website: http://tgsv5.nws.noaa.gov/tdl/. The NOAA technical report *Sea, Lake, and Overland Surges from Hurricanes* contains documentation about the TDL storm surge model called SLOSH.

Naval Research Laboratory (NRL)
Marine Meteorology Division
7 Grace Hopper Ave.
Monterey, CA 93943-5502
Phone: (408) 656-4721
Fax: (408) 656-4314
http://www.nrlmry.navy.mil/

The Naval Research Laboratory (NRL) conducts a broadly based multidisciplinary program of scientific research and advanced technological development directed toward maritime applications of new and improved materials, techniques, equipment, systems, and ocean, atmospheric, and space sciences and related technologies. NRL was commissioned in 1923 by Congress for the Department of the Navy. Today it is a field command under the Chief of Naval Research and has approximately 3,600 personnel (over 1,900 research staff, nearly half of whom hold Ph.D.s) who address basic research issues concerning the Navy's environment of sea, sky, and space. NRL's parent organization is the Office of Naval Research (ONR) in Arlington, Virginia. ONR coordinates, executes, and promotes the science and technology programs of the United States Navy and Marine Corps through universities, government laboratories, and nonprofit and for-profit organizations.

NRL has a number of major sites and facilities. The largest facility is located at the Stennis Space Center in Bay St. Louis, Mississippi. Others include a facility at the Naval Postgraduate School in Monterey, California, and the Chesapeake Bay Detachment in Maryland. Additional sites are located in Maryland, Virginia, Alabama, and Florida. The unit responsible for atmospheric research is the Marine Meteorology Division in Monterey.

Publications: NRL publishes an annual review that contains short articles covering a variety of research areas. NRL also publishes *The NRL Fact Book* every two years, which is primarily a reference source for information on NRL's organizational structure, capabilities, and points of contact. On the NRL website under the "Projects" link, worldwide satellite images of hurricanes and other weather phenomenon are displayed in a variety of formats, some of which are still experimental. Hurricane forecasters from around the world use this state-of-the-art website.

NOAA Aircraft Operations Center (AOC)

P.O. Box 6829
MacDill Air Force Base, FL 33608-0829
Phone: (813) 828-3310
Fax: (813) 828-3266
http://www.nc.noaa.gov/aoc.html

The Aircraft Operations Center (AOC) provides the Hurricane Research Division with aircraft equipped with the scientific instruments required for a hurricane research mission. The AOC was created in 1983 to consolidate the aviation assets operated by NOAA. The aircraft (1) collect data essential to NOAA's weather research on hurricanes (by the Hurricane Research Division), tornadoes, winter storms, air pollution, and other atmospheric issues; (2) provide aerial support of coastal and aeronautical charting; and (3) do aerial surveys for hydrologic research and marine mammal population prediction. The AOC pilots are members of the NOAA Corps and are the only pilots in the world who are trained and qualified to fly into hurricanes at dangerously low altitudes. These officers also frequently serve as chief scientists on program missions.

The aircraft used for hurricane research include the Lockheed WP-3D Orions and a G-IV Gulfstream jet. A description of each aircraft follows: The P-3s are the flagship aircraft of AOC and are among the most advanced airborne environmental research platforms flying today. Built in 1975, their cruising speed is 374 mph with a maximum range varying from 2,559 to 3,795 miles depending on the altitude at which they are flying. Because these are propeller-driven planes, the maximum altitude these aircraft can reach is 25,000 feet. They can carry up to 18 personnel (including 9 scientists or observers). The following weather parameters can be measured in these aircraft: (1) rainfall and cloud distribution; (2) three-dimensional wind field; (3) atmospheric electrification; (4) cloud particles and aerosols; (4) atmospheric radiation; (5) air temperature; (6) relative humidity; (7) vertical profile of temperature, relative humidity, and wind using a dropsonde; (8) surface pressure; and (9) remote sensing of sea-surface conditions.

In 1997 the Gulfstream IV was added to AOC's fleet. Because the Gulfstream is a jet, it can reach altitudes of 45,000 feet; it is also faster than the P-3s (with a cruising speed of 506 mph) and can fly further (with a maximum range of 4,687 miles). The jet's primary mission is hurricane surveillance, and it complements the work of the P-3s by flying into the steering currents of hurricanes up to 45,000 feet. The objective is to release multiple dropsondes from 40,000 feet around hurricanes, which will relay a better picture of the atmospheric conditions that surround and steer a hurricane. These data may improve hurricane track forecasts by about 20 percent. These data are also expected to improve scientific understanding of hurricane intensification.

The planes are flown by members of the NOAA Corps, the smallest of the seven uniformed services of the United States. As of the late 1990s, the NOAA Corps consisted of 299 officers with at least a bachelor of science in engineering, mathematics, physics, computer science, meteorology, oceanography, biology, or a related discipline. Officers operate NOAA ships, fly NOAA aircraft, lead mobile field parties, conduct diving operations, manage research projects, and serve in staff positions.

Publications: The AOC publishes a brochure titled *Aircraft Operations Center.* NOAA Public Affairs also publishes a series of brochures known as the *NOAA Backgrounder,* which include the following titles: *NOAA's "Hurricane Hunter" Aircraft; The Gulfstream-IV Jet: NOAA's Flying Meteorological Platform;* and *Office of NOAA Corps Operations.* The AOC website is also a good source of information. See the section "National Weather Service" for a detailed listing of other related NOAA pamphlets. Additional information about the AOC aircraft can be obtained from the Hurricane Research Division's website: http://www.aoml.noaa.gov/hrd.

Tropical Prediction Center
11691 S.W. 17th Street
Miami, FL 33165-2149
Phone: (305) 229-4470
http://www.nhc.noaa.gov

TPC is located on the campus of Florida International University in Southwestern Dade County in Miami, Florida. TPC has a public affairs officer who can be contacted at: (305) 229-4404.

The mission of the Tropical Prediction Center (TPC) is to save lives and protect property by issuing watches, warnings, forecasts, and analyses of hazardous weather conditions in the tropics. To fulfill this mission, TPC oversees the following branches:

National Hurricane Center (NHC)
NHC maintains a continuous watch on hurricanes over the Atlantic, Caribbean, Gulf of Mexico, and the Eastern Pacific from May 15 through November 30. NHC prepares and distributes hurricane watches and warnings for the general public and also prepares and distributes marine and military advisories for other users. During other parts of the year, NHC provides training for U.S. emergency managers and representatives from many other countries, improve hurricane forecasting techniques, and participates in public awareness programs.

The Tropical Analysis and Forecast Branch (TAFB)
TAFB provides forecasts every day of the year, involving marine forecasting, aviation forecasts and warnings, and surface analysis. The unit also provides satellite interpretation and satellite rainfall estimates for the international community. In addition, TAFB provides hurricane intensity estimates from the Dvorak technique and support to NHC when additional forecasters are necessary.

The Technical Support Branch (TSB)
TSB provides support for TPC computer and communications systems, including the McIDAS satellite data processing systems. TSB also maintains a small applied research unit that develops tools for hurricane and tropical weather analysis and prediction. TSB also has a storm surge group that provides information for developing evacuation procedures for coastal areas and an oceanography unit that produces ocean current and sea surface temperature analysis.

Publications: Numerous brochures are available upon request, including a hurricane preparedness pamphlet titled *What About Shutters?* A popular technical report written by three NHC forecasters, titled *The Deadliest, Costliest, and Most Intense United States Hurricanes of This Century (and Other Frequently Requested Hurricane Facts)* is available from the National Technical Information Service (NTIS) and an abbreviated but updated version of this report is available on the TPC website. Each year NHC forecasters publish annual hurricane summaries in *Weatherwise,* the *Mariner's Weather Log,* and *Monthly Weather Review* (see Chapter 6 for details on these publications). See the section "National Weather Service" for a detailed listing of other related NOAA pamphlets.

The Weather Channel
e-mail: metnet@1weather.com
http://www.weather.com

Started in 1982, the Weather Channel has been the only cable television media outlet providing 24-hour weather coverage and special weather programs seven days a week. It also provides weather forecasts to some newspapers and radio stations. During the hurricane season, a "Tropical Update" may be seen about 10 minutes before each hour. The Tropical Update shows satellite photos of all tropical disturbances, tropical storms, and hurricanes in the Atlantic and East Pacific; the western North Pacific also is sometimes shown. When hurricane landfall is possible, coverage is expanded, with frequent discussions with the Weather Channel's "hurricane experts," as well as live reports from meteorologists near the expected area of landfall.

Publications: A book titled *Hurricanes!* is available for purchase from their website or through a bookstore. Many maps and forecasts, as well as weather-related news, educational material, and seasonal features, are available on its website. The Weather Channel also airs "The Weather Classroom" Mondays and Thursdays at 4:00 through 4:30 A.M. Eastern Time. Each episode includes three separate weather-related topics lasting eight minutes each. Educators can tape the program and use it in their classrooms.

World Meteorological Organization (WMO)
Information and Public Affairs Office
41, Avenue Giuseppe-Motta
1211 Geneva 2
Switzerland
Phone: (041 22) 730-8314 or (041 22) 730-8315

Fax: (041 22) 733-2829
E-mail: ipa@www.wmo.ch
http://www.wmo.ch

World Meteorological Organization
Publications Sales Unit
Case Postale 2300
CH-1211 Geneva 2
Switzerland
Phone: (041 22) 730 83 14 or (041 22) 730 83 15
Fax: (41 22) 733 08 54
E-mail: PubSales@gateway.wmo.ch

American Meteorological Society
Attn: WMO Publications Center
45 Beacon Street
Boston, MA 02108
Phone: (617) 227-2425
Fax: (617) 742-8718
E-mail: wmopubs@ametsoc.org

The World Meteorological Organization (WMO) is the successor to the International Meteorological Organization (IMO), founded in 1873 to coordinate global research on weather and climate. In 1947, the IMO met in Washington, D.C., and decided to restructure as a specialized agency of the United Nations. This decision established the new World Meteorological Organization, which commenced operations in 1951.

The purposes of the WMO are to facilitate international cooperation in the establishment of networks of stations for making meteorological, hydrological, and other observations and to promote the rapid exchange of meteorological information, the standardization of meteorological observations, and the uniform publication of observations and statistics. It also furthers the application of meteorology to aviation, shipping, water problems, agriculture, and other human activities; promotes operational hydrology; and encourages research and training in meteorology. The Secretariat, headed by the secretary-general, is located in Geneva and serves as the administrative, documentation, and information center of the WMO. The 185-member organization provides the authoritative scientific voice on the state and behavior of the earth's atmosphere and climate.

One important function of the WMO is the World Weather Watch (WWW), whose mission is to coordinate and monitor all observational and telecommunication facilities worldwide so that every country has available the information it needs to provide weather services on a day-to-day basis. The WWW implements standardization of measuring methods and techniques, common telecommunication procedures, and the presentation of observed data and processed information in a manner that is understood by all, regardless of language.

The WWW is also involved in long-term planning and research. One component of the WWW is the Tropical Cyclone Programme. The main purpose of the Tropical Cyclone Programme is to assist WMO members, through an internationally coordinated program, in their efforts to mitigate tropical cyclone disasters. Under the coordination of the World Weather Watch, this program is designed to assist more than 50 countries in areas vulnerable to tropical cyclones, to minimize destruction and loss of life by improving forecasting and warning systems, and to encourage members to establish national disaster prevention and preparedness measures. Another contribution from this program has been several important technical publications.

Publications: The WMO produces many publications. With respect to hurricanes, the following technical reports are strongly recommended: *Global Perspectives on Tropical Cyclones* and *Global Guide to Tropical Cyclone Forecasting.* Other recent publications include *Frequency of Tropical Cyclones in the South Pacific during the Last 100 Years, and an Analysis of any Changes in these Factors; Fifteen Years of Progress and Achievement of the WMO Tropical Cyclone Programme (1980–1994); Papers Presented at the Workshop on Storm Surges for the Bay of Bengal; Regional Association IV (North and Central America) Hurricane Operational Plan 1979. New Edition 1994; Tropical Cyclones and Their Forecasting and Warning System in the North Indian Ocean;* and *Tropical Cyclone Warning Systems.* The organization has many other publications dating back to 1980. The WMO also publishes a newsletter every two months titled *Operational Newsletter,* which provides World Weather Watch centers a summary of the latest operational information; it is also available at the website under the World Weather Watch publication section. Meteorologists may also get the latest documentation and changes in WMO code, radiosonde sites, and so on, in the "Operational Publications" section of the website.

Print 6
Resources

This chapter contains a number of technical and popular books on hurricanes, separated by topical sections, as well as a few informative websites. These books range from elementary, to general, to technical in nature and focus on such topics as individual storms, hurricane preparedness, and the societal impact of hurricanes. ISBNs are also given, when possible. Many books can be ordered over the Internet directly from the publisher, from local bookstores, or from online bookstores. The American Booksellers Association (located at http://www.bookweb.org) also provides a useful directory of its member bookstores, many of whom offer online ordering services. In addition, Yahoo (at http://www.yahoo.com) contains a comprehensive list of online bookstores.

Almost all recent (after 1990) U.S. government publications and technical reports can be ordered through the National Technical Information Service (see information at the end of the chapter). Other books are out of print but are available in some libraries. Should your library not contain a particular book, try having your librarian order it through an interlibrary loan from another library. A good starting point is the NOAA library, which contains a user-friendly interface for listing books by author, subject, title, or keyword on their website (see http://www.lib.noaa.gov for details; click on its "library catalog" section to begin the reference search).

Another good source of information about specific hurricanes is newspaper accounts. Many libraries contain microfilm copies that contain fascinating stories, interviews, and spectacular pictures. For example, to learn more about the New England Hurricane of 1938, look at issues of the *New York Times* beginning September 31, 1938.

Young Adult (Elementary) Books on Hurricanes

These books contain basic information about hurricanes supplemented with good pictures. They are ideal for students and other readers looking for a short, easy-to-read book on hurricanes. The author Sally Lee also provides some science projects.

Carpenter, S. M., and T. G. Carpenter. *The Hurricane Handbook: A Practical Guide for Residents of the Hurricane Belt.* Lake Buena Vista, FL: Tailored Tours Publications, 1993. 128 pp. ISBN 0-963-12414-5.

Hood, S., and J. Barkan. *Hurricanes!* New York: Simon & Schuster Children's, 1998. 64 pp. ISBN 0-689-82017-8.

Lauber, P. *Hurricanes—Earth's Mightiest Storms.* New York: Scholastic Press, 1996. 64 pp. ISBN 0-590-47406-5.

Lee, S. *Hurricanes.* New York: Franklin Watts, 1993. 63 pp. ISBN 0-531-15665-6.

General Books on Hurricanes

Allaby, M. *A Chronology of Weather.* New York: Facts on File, 1998. 154 pp. ISBN 0-8160-5321-0.

The largest part of this book consists of two chronological accounts. One lists some of the major weather disasters (e.g., hurricanes, tornadoes, droughts, etc.) by the years in which they occurred, starting in 3200 B.C. and continuing through 1997. The other chronological account lists important developments in the understanding of weather and improvements in forecasting. This book also contains 30 simple weather experiments that can be done at home to elucidate basic weather principles.

Chaston, P. R. *Hurricanes!* Kearney, MO: Chaston Scientific, 1996. 182 pp. ISBN 0-9645172-2-1.

This book, written by a meteorologist, covers many aspects of hurricanes, including hurricane life cycles, hurricane structure, the Saffir-Simpson scale, wind destruction and storm surge damage, spectacular hurricane stories from survivors, and so on, including 100 photos from reconnaissance flights, weather satellites, and radars as well as easy-to-understand graphics. The history of hurricanes is chronicled from Columbus's time through today. Maps detail the tracks of hurricanes, and the naming of these storms is explained. The hurricane category scale of intensity is explained.

Cline, I. M. *Tropical Cyclones.* New York, Macmillan Company, 1926. 301 pp. ISBN 0-9645172-2-1.

When it was written, this was the most authoritative book in the United States on hurricanes. It contains considerable original work on hurricane rainfall, tides, and waves, with the focus on Gulf of Mexico storms. This book is out of print.

Douglas, M. S. *Hurricane.* Atlanta: Mockingbird Books, 1976. 119 pp. ISBN 89176-015-6.

The author is the well-known Florida Everglades activist and writer who recently passed away at the age of 107. Originally written in 1958, the first half is an interesting perspective on hurricane knowledge in that presatellite era. The second half of the book contains a detailed history of how weather forecasting evolved in the United States from the time of Benjamin Franklin to the Galveston Hurricane of 1900.

Dunn, G. E., and B. I. Miller. *Atlantic Hurricanes.* Baton Rouge: Louisiana State University Press, 1960. 377 pp.

When this book was written, Gordon Dunn was the director of the National Hurricane Center, and Banner Miller was a research meteorologist at NHC. At the time of its publication, it was the most comprehensive book on hurricanes and an amazing accomplishment when one considers that many of the conclusions were based on ship, radar, tide, and aircraft data (at that time, no buoy or satellite observations were available). It still is an excellent book for detailed reading on hurricanes, although advancements in the last 40 years have made some parts obsolete. The book also contains an excellent chronology of events affecting hurricane research and forecasting, as well as biographies of pertinent early hurricane scientists. This book is out of print.

Fisher, D. E. *The Scariest Place on Earth.* New York: Random House, 1994. 250 pp. ISBN 0-679-42775-9.

David Fisher is a professor of cosmochemistry and director of the Environmental Science Program at the University of Miami, and he personally experienced Hurricane Andrew's landfall. This book is written in a three-topic fashion, jumping back and forth between topics. These three topics cover historical accounts and facts on hurricanes, a description of hurricanes, and a chronological account of Fisher's Andrew experience. Fisher also presents a reasonably balanced discussion on whether or not global warming will increase the number and intensity of hurricanes. However, some readers may not appreciate his conclusion that global warming is already occurring.

Fishman, J., and R. Kalish. *The Weather Revolution—Innovations and Imminent Breakthroughs in Accurate Forecasting.* New York: Plenum Press, 1994. 276 pp. ISBN 0-306-44764-9.

This excellent book discusses the history of weather-related technological advances, including computer models, satellites, hurricane forecasting, severe weather forecasting, and long-term weather prediction. The authors write about complex topics in a thorough but understandable fashion.

Fleming, J. R., ed. *Historical Essays on Meteorology.* Boston: American Meteorological Society, 1996. 617 pp. ISBN 1-878220-17-9.

This book chronicles achievements in theoretical meteorology, computer models, observation tools, cloud physics, hurricanes, convection, lightning, climatology, hydrology, the private sector, and education. Mark DeMaria, a former researcher at the National Hurricane Center and the Hurricane Research Division, presents a history of hurricane forecasting from 1920 to the late 1990s in Chapter 9.

Ludlum, D. M. *Early American Hurricanes: 1492–1870.* Boston: American Meteorological Society, 1989. 198 pp. ISBN 0-933876-16-5.

This book describes, in chronological order, the occurrences of hurricanes prior to 1870 that either closely approached or actually crossed the Atlantic and Gulf coastlines of the United States. The book is divided into two periods. The first section starts with the voyages of Christopher Columbus, extends through the years of exploration and colonization, and ends in 1814. This year makes a convenient break, for the first federal attempts to institute a national weather observing service started during the War of 1812. With the cessation of hostilities in 1815, an era of worldwide peace emerged, with vigorous worldwide trading accompanied by merchant marine weather logs. The second section spans this period, carries through the Civil War to 1870, the year the U.S. Signal Corps established its storm warning system. This book's documentation is meticulous, and a copy is a must for any weather historian.

Simpson, R. H., and H. Riehl. *The Hurricane and Its Impact.* Baton Rouge: Louisiana State University Press, 1981. 398 pp. ISBN 0-8071-0688-7.

Written by the former director of NHC (Simpson) and by the "father of tropical meteorology" (Riehl), this 1981 manuscript is a follow-up to Dunn and Miller's 1960 book. Because of the tremendous observational and research improvements since 1960, this book is more representative of current understanding of hurricanes. It is more scientific in scope than other books in this section, but still quite readable for most people. This book is out of print.

Tannehill, I. R. *Hurricanes—Their Nature and History,* rev. ed. Princeton: Princeton University Press, 1956. 308 pp.

Ivan Tannehill had just retired as chief of the Division of Synoptic Reports and Forecasts of the U.S. Weather Bureau when he wrote the original book in 1938. This 1956 version is an update to the original. Tannehill had spent 20 years engaged in hurricane warning work. The first half discusses the knowledge of hurricanes at that time, their frequency, and how to know a hurricane was coming by watching storm swells, wind direction, and pressure readings. (Recall that there were no satellites at this time, and hurricanes still surprised sailors and coastal residents.) Several unique pictures are also included, such as dozens of birds caught in a hurricane resting on a ship also trapped in the eye of a hurricane. The second half is a detailed historical perspective of Atlantic hurricanes from the 1400s to 1955. This book is out of print.

Frequently Asked Questions about Hurricanes

Tuffy, B. *1001 Questions Answered about Hurricanes, Tornadoes and Other Natural Air Disasters.* Toronto, Ontario: Dover Publications, 1987. 381 pp. ISBN 0-486-25455-0.

As suggested in the title, this is a book answering frequently asked questions about meteorology. Topics include hurricanes, tornadoes, thunderstorms, hailstorms, wind, fog, snowstorms, and ice storms. Weather records are listed, although some of these records have been broken since publication. For example, the book still cites Hurricane Betsy as causing the most property damage. Today Betsy is eleventh (not considering inflation). Also included are 72 drawings, 20 photographs, and eyewitness accounts.

Popular Weather Magazines

The *Mariner's Weather Log,* a publication of the National Weather Service (NWS), contains articles, news, and information about marine weather events and phenomena, storms at sea, weather forecasting, the NWS Voluntary Observing Ship (VOS) Program, Port Meteorological Officers (PMOs), cooperating ships officers, and their vessels. It provides meteorological information to the maritime community and contains a comprehensive chronicle on marine weather. It recognizes ships officers for their efforts as voluntary weather observers and allows the NWS to maintain contact with and communicate with over 10,000 shipboard observers (ships officers) in the merchant marine, NOAA Corps, Coast Guard, Navy, and so on. It also provides annual summaries of hurricane activity for all ocean basins worldwide.

The *Mariner's Weather Log* is published three times yearly, in April, August, and December of each year. It has a circulation of 8,100, with copies distributed to mariners, marine institutions and the shipping industry, scientists and interested nonscientists, educational and research facilities, educators, libraries, and government agencies and offices worldwide.

Weatherwise, available bimonthly, is a popular magazine intended for general public weather enthusiasts. Articles are written about a wide variety of weather-related issues, phenomena, and history, including hurricanes. In the February/March issue there are articles describing the Atlantic and Pacific hurricane seasons from the preceding year. Compelling columns and color photographs are also included.

Hurricane Preparation

Staff of the *Miami Herald. Hurricanes—How to Prepare & Recover.* Kansas City, MO: Andrews and McMeel, 1993. 125 pp. ISBN 0-8362-1718-7.

Based on the Hurricane Andrew experience, this book provides in-depth instructions on how to prepare one's home and family for a hurricane onslaught. Instructions for recovery and repairs are also included.

Societal Impact of Hurricanes

Cook, R. A., and M. Soltani, eds. *Hurricanes of 1992.* New York: American Society of Civil Engineers, 1994. 808 pp. ISBN 0-7844-0046-6.

Three major hurricanes hit U.S. coasts or territories in 1992—Hurricane Andrew, Hurricane Iniki, and Typhoon Omar—causing over $35 billion in damages. This book contains papers presented at a conference in Miami to disseminate information on wind hazards and managing wind-related disasters based on experiences with these three tropical cyclones. Topics include (1) wind speeds and wind loads; (2) risk assessment; (3) insurance; (4) damage assessment; (5) building codes; (6) building code implementation and enforcement; (7) coastal structures; (8) manufactured, residential, and commercial structures; (9) essential facilities; and (10) lifelines.

Diaz, H. F., ed. *Hurricanes—Climate and Socioeconomic Impacts.* Verlag, Germany: Springer, 1997. 292 pp. ISBN 3-540-62078-8.

This book contains reports from a workshop titled "Atlantic Hurricane Variability on Decadal Time Scales: Nature, Causes, and Socioeconomic Impacts" that was held at the National Hurricane Center on February 9–10, 1995. Climate change issues are discussed along with such socioeconomic issues as private insurance losses and the availability of insurance for coastal residents in the future. This workshop assembled experts with different views on the range and predictability of potential climatic changes and their impacts on tropical cyclone activity. The potential of massive economic losses when Atlantic hurricane activity increases to the level observed in the 1940s and 1950s is discussed.

Elsner, J. B., and A. Birol Kara. *Hurricanes of the North Atlantic.* New York: Oxford University Press, 1999. 496 pp. ISBN 0-19-512508-8.

This book is intended for both climate researchers and users of hurricane data. The book begins with a general description of hurricanes and provides statistical information regarding their tracks, origins, intensity, and duration. The authors then present similar statistical information for hurricanes divided into three categories: 1) hurricanes that had only tropical influences; 2) hurricanes that had nontropical influences, such as interaction with a cold front; and 3) major hurricanes. Statistical information is then presented on landfalling hurricanes in the United States, Puerto Rico, Jamaica, and Bermuda. An analysis on hurricane climate cycles, trends, and return periods is then presented. The authors give a history of seasonal hurricane forecasting and describe seasonal forecasting schemes currently being used. Finally, the authors discuss society's vulnerability to hurricanes. Statistics on coastal population change and property development are presented, and ideas on risk management and catastrophe insurance are discussed.

Pielke, R. A., Jr., and R. A. Pielke Sr. *Hurricanes: Their Nature and Impacts on Society.* Chichester, England: John Wiley & Sons, 1997. 279 pp. ISBN 0-471-97354-8.

This book discusses the increased vulnerability of the U.S. coastline to hurricanes due to population increases and property development. It addresses both the scientific and societal impacts of hurricanes and illustrates the economic benefit of hurricane research in an era in which scientific research is under pressure to demonstrate a better connection to societal needs. Suggestions for policy implementation are included. A well-done description of hurricanes is contained in the first few chapters.

A good supplementary text, which includes a more basic description of hurricanes, archives of annual Atlantic hurricane tracks, and pictures of the ocean inside hurricanes of different intensities, is *The Hurricane,* by Roger A. Pielke (London: Routledge 1990).

Technical Books

Anthes, R. A. *Tropical Cyclones—Their Evolution, Structure, and Effects.* Boston: American Meteorological Society, 1982. 208 pp. ISBN 0-933876-54-8.

Although it contains somewhat dated material, this book contains interesting sections discussing the mathematical framework of hurricanes, their structure, and life cycle. Another chapter nicely summarizes the history of hurricane computer simulations. The reader should be warned that this book is quite technical, but it gives a flavor of what true hurricane research is about.

Elsberry, R. L., W. M. Frank, G. J. Holland, J. D. Jarrell, and R. L. Southern. *A Global View of Tropical Cyclones.* Monterey, CA: Naval Postgraduate School, 1987. 195 pp. Funded by the Office of Naval Research.

This book results from the International Workshop on Tropical Cyclones held in Bangkok, Thailand, from November 25 to December 5, 1987. This workshop brought together 83 experts in tropical cyclone forecasting, research, and warning strategies from 28 different countries and was the first truly worldwide gathering of tropical cyclone specialists. The result is a book that many of the leading tropical cyclone experts still reference in their work. Some parts are technical, but most of the book is quite readable. Chapters are included on tropical cyclone observations, structure, genesis, motion, and their impacts. The final chapter discusses warning and mitigation strategies.

The book is now out of print, but copies are available at many university libraries, the Hurricane Research Division, the National Hurricane Center, the Naval Postgraduate School, and the Office of Naval Research.

Foley, G. R., H. E. Willoughby, J. L. McBride, R. L. Elsberry, I. Ginis, and L. Chen. *Global Perspectives on Tropical Cyclones.* Report No. TCP-38. Geneva, Switzerland: World Meteorological Organization, 1995. 289 pp.

In many respects, this book is a revision of Elsberry et al., *A Global View of Tropical Cyclones,* and reflects the many developments in tropical cyclone research since 1987. This book describes the state of hurricane science through 1994. An additional chapter on ocean interactions with tropical cyclones has also been added.

This book compliments the World Meteorological Organization's operational forecasting publication *A Global Guide to Tropical Cyclone Forecasting,* by G. J. Holland.

Hebert, P. J., J. D. Jarrell, and M. Mayfield. *The Deadliest, Costliest, and Most Intense United States Hurricanes of This Century (and Other Frequently Requested Hurricane Facts).* NOAA Technical Memorandum NWS TPC-1, National Oceanic and Atmospheric Administration, National Weather Service, National Hurricane Center, Miami, 1997.

As suggested by the title, this technical report provides detailed data about the most destructive U.S. hurricanes of the twentieth century. Statistics are included about the most expensive hurricanes, the most intense hurricanes, U.S. landfalls by state and by decade, and other hurricane facts.

Holland, G. J, 1993: *Global Guide to Tropical Cyclone Forecasting.* Report No. TCP-31. Geneva, Switzerland: World Meteorological Organization, 1993. 347 pp.

The purpose of this book is to compliment the theoretical and descriptive content of *Global Perspectives on Tropical Cyclones* and *A Global View of Tropical Cyclones* with a practical guide to the applied aspects of tropical cyclone forecasting. Included are chapters on track forecasting, satellite interpretation, storm surge forecasting, seasonal forecasting, forecast center strategies, and warning procedures. An informative appendix of tables, definitions, and tropical cyclone records is included at the back of the book.

Rappaport, E. N., and J. Fernandez-Partagas. *The Deadliest Atlantic Tropical Cyclones, 1492–1994.* NOAA Technical Memorandum NWS NHC 47, National Oceanic and Atmospheric Administration, National Weather Service, National Hurricane Center, Miami, 1995.

This technical report contains information on tropical cyclones that have caused 25 or more deaths in any part of the Atlantic basin. It also contains information on cyclones that may have caused 25 or more deaths, but where the total was not quantified by sources. Updated versions of this report are contained on the National Hurricane Center website at http://www.nhc.noaa.gov.

General Meteorology Books and Textbooks

Hurricanes embody almost all meteorological processes. For an understanding of hurricanes, a solid background in all atmospheric phenomena is required. Therefore, the interested reader is urged to consult a good general meteorology textbook. A few of the many possible choices are listed in this section.

Ahrens, C. D. *Meteorology Today.* St. Paul, MN: West Publishing Co., 1994. 591 pp. ISBN 0-314-02779-3.

Danielson, E. W., J. Levin, and E. Abrams. *Meteorology.* Boston: WCB/McGraw-Hill Co., 1998. 462 pp. ISBN 0-697-21711-6.

Nese, J. M., and L. M. Grenci. *A World of Weather: Fundamentals of Meteorology.* Dubuque, IA: Kendall/Hunt Publishing Co., 1998. 539 pp. ISBN 0-7872-3578-4.

Williams, J. *The Weather Book.* Arlington, VA: USA Today, 1992. 212 pp. ISBN 0-679-73669-7.

Encyclopedias and Glossaries on Meteorology

Geer, Ira W., ed. *Glossary of Weather and Climate*. Boston: American Meteorological Society, 1996. 272 pp. ISBN 1-878220-21-7.

Written for a general audience, this book contains a glossary of over 3,000 terms frequently used in discussions and descriptions of meteorological and climatological phenomena. In addition, the glossary includes definitions of related oceanic and hydrologic terms. Currently, the AMS is revising and expanding this book (with hundreds of volunteers) so that it will include over 13,000 terms. It is anticipated that this revised edition will be published in 1999 and that it will become available on the AMS website, http://www.ametsoc.org/AMS.

Schneider, S. H. *Encyclopedia of Climate and Weather*. New York: Oxford University Press, 1998. 929 pp. ISBN 0-19-509485-9.

In addition to consulting a good general meteorology textbook, a meteorology encyclopedia is useful because hundreds of topics are arranged in alphabetical order with liberal cross-references. This two-volume set is suitable for high school and above, with clear explanations and abundant figures on a variety of climate and weather topics. Each article is written by an authority on the subject. A glossary is provided in the back.

Regional Books

Barnes, Jay. *Florida's Hurricane History*. Chapel Hill: The University of North Carolina Press, 1998. 384 pp. ISBN 0-8078-4748-8.

This book, based on newspaper reports, National Weather Service records, books, and eyewitness accounts, provides a detailed record of over 100 hurricanes that have struck Florida in the last 450 years. Information on the basics of hurricane structure, formation, naming, and forecasting is included, as well as over 200 photographs, maps, and illustrations.

Barnes, Jay. *North Carolina's Hurricane History*. Chapel Hill: The University of North Carolina Press, 1998. 206 pp. ISBN 0-8078-4728-3.

This book, based on newspaper reports, National Weather Service records, books, and eyewitness accounts, provides a detailed record of more than 50 hurricanes that have struck North Carolina from 1526 to 1996. A chapter is also included on northeasters, which are nontropical winter cyclones that are near hurricane intensity; northeasters inflict storm surge and wind damage similar to tropical storms and Category 1 hurricanes.

Stevenson, J. D. *A Historical Account of Tropical Cyclones That Have Impacted North Carolina since 1586.* Bohemia, NY: National Weather Service, Scientific Services Division, Eastern Region Headquarters, 1990.

This book contains a chronology of tropical storms and hurricanes that have hit the North Carolina area from 1586 to the 1980s.

Sullivan, C. L. *Hurricanes of the Mississippi Gulf Coast, 1717 to Present.* Biloxi, MS: Gulf Publishing, 1986. 139 pp. ISBN 0-8078-4507-8.

This book contains detailed information about individual tropical storms and hurricanes that have hit Mississippi from 1717 to the mid-1980s. It is out of print.

Vallee, David, and M. R. Dion. *Southern New England Tropical Storms and Hurricanes, A Ninety-Eight Year Summary 1909–1997.* Taunton, MA: National Weather Service, Scientific Services Division, Eastern Region Headquarters, 1998.

This book contains detailed information about individual tropical storms and hurricanes that have hit the New England area in the twentieth century.

Williams, John M, and Iver W. Duedall. *Florida Hurricanes and Tropical Storms.* Gainesville: University Press of Florida, 1997. 148 pp. ISBN 0-8130-1515-4.

This book is a comprehensive chronological guide to hurricanes, tropical storms, and near-misses that have affected Florida since 1871. Additional features include statistics for each hurricane and tropical storm, eyewitness accounts, photos, 10-year tracking charts, and a hurricane preparedness checklist.

Individual Hurricanes

The Galveston Hurricane of 1900

Weems, J. E. *A Weekend in September.* College Station: Texas A & M University Press, 1997. 192 pp. ISBN 0-89096-390-8.

Weems interviews survivors of the worst natural disaster in U.S. history—the Galveston Hurricane of 1900, which inundated the island with an 8-to-15-foot storm surge that killed at least 6,000 people on the island and 2,000 further inland. Efforts to protect Galveston from future hurricanes are also discussed. A six-mile-wide seawall was completed in 1905, and the elevation of the entire city was raised 8 feet, to a height of 17 feet. A storm in 1915 tested the newly reshaped island, and although there was some damage, the community survived. Based on this result, the width of the seawall was expanded to 10.5 miles.

Larson, E. *Isaac's Storm.* New York: Crown Publishers, Random House, 1999. 323 pp. ISBN 0-609-60233-0

This book is a combined dramatization/factual account of the 1900 Galveston Hurricane, with the island's chief meteorologist Isaac Cline as the central character. This book delves into Cline's career and character and exposes the ineptitude, corruption, and arrogance of Weather Bureau administrators in Washington, DC (who ignored warnings from Cuban meteorologists that a hurricane was in the Gulf of Mexico and instead stated that the storm was headed into the Atlantic; no warning was ever issued from Washington, DC). This is probably the most factual book on Isaac Cline and the Galveston hurricane to date and exposes many myths about this tragedy. Also included is brief historical information about other Atlantic hurricanes before 1900.

The 1926 Florida Hurricane

Reardon, L. F. *Florida Hurricane and Disaster, 1926.* Coral Gables, FL: Arva Parks & Company, 1992. 112 pp. ISBN 0-914381-04-0.

This is a reproduction of a survivor's diary of the 1926 hurricane that hit Miami and killed 243 people. The text has been supplemented with pictures. It also includes a description of Hurricane

Andrew (1992), which is uniquely appended upside down starting on the back cover.

The 1928 Okeechobee Hurricane

Will, L. E. *Okeechobee Hurricane—Killer Storms in the Everglades.* St. Petersburg, FL: Great Outdoors Publishing Co., 1971. 204 pp.

This book is an eyewitness account of the 1928 Okeechobee Hurricane, which killed 1,836 people and injured 1,849. Most of the deaths were by drowning, but some were caused by snakebites as people climbed into trees to escape the flood waters, only to find hordes of venomous water moccasins also seeking shelter. Although the lake is actually inland, the hurricane forced a storm surge that broke the eastern earthen dike on the southern end of Lake Okeechobee. This calamity occurred within a few miles of a large city and a world famous resort, yet so isolated was the location that no one knew what had happened for three days afterward. The author also discusses a similar situation from another hurricane that occurred two years earlier and killed about 200 at the Okeechobee town of Moore Haven. Pictures of the aftermath from both of these hurricanes are included. As a result of these disasters, a levee (called the Herbert Hoover Dike in honor of the president who supported its construction) that compares in size to the Great Wall of China was built to prevent further disasters.

The New England Hurricane (1938)

Allen, E. S. *A Wind to Shake the World: The Story of the 1938 Hurricane.* Boston: Little, Brown, and Co., 1976. 370 pp.

Allen had just started his first job as a newspaper reporter when the storm hit September 21, 1938. The book recounts his own experiences and the stories of others caught by the storm. This book is out of print.

McCarthy, J. *Hurricane!* New York: American Heritage Publishing Co., 1969. 168 pp. ISBN 8281-0020-9.

This book provides a detailed account of the 1938 New England Hurricane, which made landfall on the south shore of Long Island, New York. This storm surprised New England residents who were unaccustomed to hurricanes (the last hurricane had hit New England in 1815). Because the hurricane moved at a fast speed of 60 mph and because the Weather Bureau had thought it would recurve offshore and not hit the United States, there was little preparation time available. Even worse, the storm hit at the equinox, when tides are usually their highest. Ten-foot waves atop a 20-foot storm surge killed 600 people and caused immense property damage. Barrier islands were swept so bare that rescue workers used phone company charts to determine where houses once stood. This book includes detailed eyewitness accounts and photos of the storm's aftermath.

Minsinger, W. E. *The 1938 Hurricane.* Boston: American Meteorological Society, 1988. 128 pp. ISBN 9-993382-16-7.

To mark the fiftieth anniversary of this destructive hurricane, the American Meteorological Society, in cooperation with the Blue Hill Meteorological Observatory, published this detailed historical volume, complete with over 120 archival photographs and weather charts.

The Great American Hurricane (1944)

The North Atlantic Hurricane. Charleston, SC: Historical Publications. 64 pp.

This storm received its name because it traveled up the Atlantic coast from North Carolina to the northeast, making landfall in Rhode Island. This publication chronicles the path and destruction of this storm, which affected the whole East Coast.

Hurricane Hazel (1954)

Hurricane Hazel Lashes North Carolina—The Great Storm in Pictures. Charleston, SC: Historical Publications.

This publication shows the damage caused by Hazel on the North Carolina shoreline. Prior to 1954, Hazel was the most destructive storm in North Carolina's history, producing wind gusts of 150 mph and a high storm surge superimposed on the highest ocean tide of the year.

Hurricane Carla (1961)

Hogan, W. L., ed. *Hurricane Carla—A Tribute to the News Media.* Houston: Leaman-Hogan Co., 1961. 192 pp.

Carla (1961), the largest Gulf of Mexico hurricane in decades, was perhaps the first hurricane that attracted massive, coordinated media attention. This book is a compilation of the original newspaper stories from many Texas media outlets describing Carla's widespread destruction and fatalities. Also included are accounts from several television and radio stations. Many pictures from various media outlets are also included. Carla also launched the career of Dan Rather, who was the news director of KHOU-TV. Doing a remote broadcast from the Galveston Weather Bureau, Rather acted as its spokesman, issuing continuous advisories, bulletins, evacuation instructions, and hurricane reports for almost four days. CBS management was impressed and made Rather a national correspondent. This book is out of print.

Hurricane Camille (1969)

Miss Camille 1969. Charleston, SC: Historical Publications.

This publication contains photographs of Hurricane Camille's (1969) incredible destruction on the Louisiana, Mississippi, and Alabama coasts and the Florida Gulf Coast. This storm was one of only two U.S. hurricanes that made landfall as Category 5 storms.

U.S. Army Corps of Engineers. *Hurricane Camille—14–22 August 1969.* Mobile, AL: U.S. Army Engineer District, 1970.

This report contains detailed meteorological and storm surge data from the Louisiana coast to the Alabama coast for Hurricane Camille (1969). Included are comprehensive storm surge maps, many pictures, and a summary of evacuation and relief activities. Tabular data and a description of storm damage are presented with regard to residential, commercial, industrial, government, and agricultural property.

Cyclone Tracy (1974)

Bunbury, B. *Cyclone Tracy, Picking up the Pieces.* South Fremantle, Australia: Fremantle Arts Centre Press, 1994. 148 pp. ISBN 1-86368-112-4.

This book is a chronological account of the 1974 Christmas Eve cyclone that devastated Darwin, Australia, and killed at least 48 people in the city and an additional 15 at sea. Eyewitness accounts and commentary are included throughout the book, with some pictures of the aftermath.

Staff of the *Sydney Morning Herald.* **Cyclone! Christmas in Darwin 1974.** Artarmon, Australia: John Sands, 1975.

This is another pictorial account of Cyclone Tracy's devastation. This book is out of print.

Hurricane Frederic (1979)

Hurricane Frederic. Charleston, SC: Historical Publications. 88 pp.

This publication contains photographs of Frederic's impact on southern Alabama. Interviews with those who experienced Frederic's landfall are included.

Hurricane Alicia (1983)

Hurricane Alicia: Thursday—August 18, 1983. Lubbock, TX: Barron Publications, 1983. 80 pp.

This is a collection of photographs of Hurricane Alicia's damage. This book is conveniently separated by location. For example, the cities of Baytown, Texas City, Houston, and Galveston are each covered in a separate section displaying the damage to their communities. Separate sections are also allocated to such smaller communities as Jamaica Beach, Terramar Beach, Surfside, Sea Isle, Hitchcock, Kemah, and so on. Articles on Alicia and some meteorological background on hurricanes are included. An article on the Galveston Hurricane of 1900 is also presented. This book is out of print.

Kareem, A., ed. **Hurricane Alicia: One Year Later.** New York: American Society of Civil Engineers, 1985. 335 pp.

Due to its inland passage over downtown Houston, Hurricane Alicia (1983) caused over a billion dollars in damage, providing a glimpse of what hurricanes could do in the future in heavily developed coastal areas. This proceedings contains papers presented at a conference in Galveston, Texas, devoted to areas of meteorology, structural behavior, hurricane-resistant design, and building codes.

Hurricane Hugo (1989)

Fox, William P. *Lunatic Wind: Surviving out the Storm of the Century.* Chapel Hill, NC: Algonquin Books of Chapel Hill, 1992. 197 pp. ISBN 0945575424.

Fox, who has experience writing fiction and filmstrips, documented the landfall of Hugo near Charleston, South Carolina, in a dramatic style. Based on detailed research and interviews, Fox recreated the night of the storm and the people caught in it. Mixing factual reporting with the storytelling of a novelist, Fox's docudrama recreates the experience of people who survived—two teenage surfers who had to swim for their lives as their beachhouse got destroyed by the storm surge; a shrimp boat captain determined to ride out the storm in the town's harbor; and over 1,000 people in an evacuation shelter (a high school gym) that was flooded by the storm surge because the shelter was at a lower elevation than local charts had indicated.

Storm of the Century—Hurricane Hugo. Charleston, SC: Historical Publications. 76 pp.

This publication contains photographs of Hugo's impact on Charleston. Interviews with those who experienced Hugo's landfall are included. One interview vividly describes a family's struggle with Hugo's storm surge as it broke through their home, causing them to scamper to the attic.

Hurricane Bob (1991)

New England's Nightmare—Hurricane Bob. Charleston, SC: Historical Publications. 97 pp.

This publication contains photographs of Bob's impact on the northeastern United States. Interviews with those who experienced Bob's landfall are included.

Hurricane Andrew (1992)

Bailey, William E. *Andrew's Legacy: Winds of Change.* West Palm Beach, FL: Telshare Publishing, 1999. ISBN 0-910-287-14-7

Dr. William E. Bailey is an attorney and consultant specializing in insurance industry issues and litigation. In October 1992, Bailey was appointed co-director of the Hurricane Insurance Information Center in Miami, Florida. The center, established right after

Hurricane Andrew wreaked its havoc on South Florida, was formed by seven national insurance trade associations and twenty-two major insurance companies representing the largest segment of property insurers in the state. Its one-year mission was to inform and educate consumers and the media regarding insurance issues in the aftermath of the costliest natural disaster in United States history. One of the results of this experience is this book, which discusses implications on possible future hurricane landfalls in metropolitan regions.

Florida's Path of Destruction—Hurricane Andrew. Charleston, SC: Historical Publications. 97 pp.

This publication contains over 300 photographs of Andrew's impact on the Bahamas, Miami, and Louisiana. Interviews with weather specialists and those who experienced its landfall are included, as are satellite photos.

Mirza, Noorina, and M. Quraishy. *Before and after Hurricane Andrew 1992.* Miami: Kenya Photo Mural, 1992. 96 pp. ISBN 0-963-4962-0-4.

This is another picture collection of Hurricane Andrew's damage. Much of the book contains interesting comparative pictures of what areas looked like before Andrew and what they looked like immediately afterward.

Staff of the *Miami Herald* and *El Nuevo Herald. Hurricane Andrew—The Big One.* Kansas City, MO: Andrews and McMeel, 1992. 160 pp. ISBN 0-8362-8012-1.

A humbling, touching collection of photographs displaying the incredible destruction of Hurricane Andrew and its impact on the people of Dade County south of Miami. It is impossible to describe these pictures in words. As one survivor says in the book, "It's something you can't talk about. You just have to see it." Also included are eyewitness accounts and reports in a chronological fashion.

U.S. Army Corps of Engineers. *Hurricane Andrew Assessment.* Tallahassee, FL: Post, Buckley, Schuh & Jernigan, Inc., 1993.

Andrew provided the opportunity for the Federal Management Agency and the Army Corps of Engineers to investigate whether state and local officials used their products, whether their past studies were useful in preparing for Andrew, and which of these past studies were most useful or least useful. This report gives their findings.

Hurricane Iniki (1992)

Hurricane Iniki. Charleston, SC: Historical Publications. 100 pp.

This publication contains photographs of Iniki's impact on Kauai in the Hawaii Islands, including unique documentation of the landfall itself because Iniki hit during the day (unlike many other hurricanes, such as Hugo and Andrew, which hit at night).

Hurricane Alberto (1994)

Deadly Waters—Tropical Storm Alberto. Charleston, SC: Historical Publications. 92 pp.

This publication contains photographs of Alberto's impact, which produced record-breaking rainfall and floods that took 28 lives in Georgia and 2 in Alabama. Interviews with those who experienced Alberto are included.

Hurricane Georges (1998)

Hurricane Georges. Charleston, SC: Historical Publications. 52 pp.

This publication contains photographs of Georges's impact on the Florida Keys as documented through the *Key West Citizen* and *Free Press* reporters who live on the Keys. The book includes interviews with those who experienced Georges's landfall, such as those who live on "houseboat row" and the popular Duval Street area.

Historical Hurricane Tracks

Algue, J. *The Cyclones of the Far East.* Manila: Philippine Weather Bureau, 1904. 283 pp.

Much hurricane knowledge was obtained from Roman Catholic priests living on oceanic islands. Rev. Algue's book presents observations of typhoons in great detail and is probably the first thorough documentation of western Pacific tropical cyclones. Temperature, wind, and pressure values are presented in a tabular form, and analyses of typhoon landfalls are described. Also included are projected tracks of late 1800 and early 1900 storms and many climatology maps. Atmospheric conditions that indicate the approach of a possible typhoon are presented. This book is out of print.

Lourensz, R. S. *Tropical Cyclones in the Australian Region, July 1909 to June 1980.* Maryborough, Victoria: Bureau of Meteorology, Department of Science and Technology, Australian Government Publishing Service, 1981. 94 pp. ISBN 0-642-01718-2.

This book presents annual tracks for cyclones in the oceanic regions near Australia from 1909 to 1980. Basic statistical summaries and graphics are also presented.

Maunder, J., 1995: *An Historic Overview Regarding the Intensity, Tracks, and Frequency of Tropical Cyclones in the South Pacific during the Last 100 Years, and an Analysis of Any Changes in These Factors.* Report No. TCP-37. Geneva, Switzerland: World Meteorological Organization, 1995. 48 pp.

As suggested by the title, this book presents annual tracks for cyclones in the South Pacific for the 1890s–1990s, as well as tabular and statistical information.

Neumann, C. J., B. R. Jarvinen, C. J. McAdie, and J. D. Elms. *Tropical Cyclones of the North Atlantic Ocean, 1871–1992.* Asheville, NC: National Environmental Satellite, Data, and Information Service, National Climatic Data Center, 1993. 193 pp.

This valuable book presents annual tracks for tropical depressions, tropical storms, and hurricanes for the Atlantic Ocean from 1871 to 1992. Also included in these tracks are an additional category of tropical cyclones called subtropical storms, which contain different temperature characteristics from hurricanes but can still produce rain and moderate winds. Basic statistical summaries are presented as well.

National Hurricane Center's Annual Hurricane Summary

Approximately once a year, the publication *Monthly Weather Review* has a summary of the previous year's hurricane activity authored by National Hurricane Center forecasters. The summaries include descriptions of named storms along with pertinent meteorological data and satellite imagery. Similar reviews are also published in the February/March issue of *Weatherwise* that are less technical and describe the Atlantic and Pacific hurricane seasons from the preceding year.

The references for *Monthly Weather Review* summaries since 1980 are:

Lawrence, M. B. 1981. Atlantic Hurricane Season of 1980. *Mon. Wea. Rev.*, 109, 1567–1582.

Lawrence, M. B. 1982. Atlantic Season of 1981. *Mon. Wea. Rev.*, 110, 852–866.

Clark, G. B. 1983. Atlantic Season of 1982. *Mon. Wea. Rev.*, 111, 1071–1079.

Case, R. A., and H. P. Gerrish. 1984. Atlantic Hurricane Season of 1983. *Mon. Wea. Rev.*, 112, 1083–1092.

Lawrence, M. B., and G. B. Clark. 1985. Atlantic Hurricane Season of 1984. *Mon. Wea. Rev.*, 113, 1228–1237.

Case, R. A. 1986. Atlantic Hurricane Season of 1985. *Mon. Wea. Rev.*, 114, 1390–1405.

Lawrence, M. B. 1988. Atlantic Hurricane Season of 1986. *Mon. Wea. Rev.*, 116, 2155–2160.

Case, R. A., and H. P. Gerrish, 1989. Atlantic Hurricane Season of 1987. *Mon. Wea. Rev.*, 117, 939–949.

Lawrence, M. B., and J. M. Gross, 1989. Atlantic Hurricane Season of 1988. *Mon. Wea. Rev.*, 117, 2248–2259.

Case, B., and M. Mayfield, 1990. Atlantic Hurricane Season of 1989. *Mon. Wea. Rev.*, 118, 1165–1177.

Mayfield, M., and M. B. Lawrence, 1991. Atlantic Hurricane Season of 1990. *Mon. Wea. Rev.*, 119, 2014–2026.

Pasch, R. J., and L. A. Avila. 1992. Atlantic Hurricane Season of 1991. *Mon. Wea. Rev.*, 120, 2671–2687.

Mayfield, M., L. A. Avila, and E. N. Rappaport, 1994. Atlantic Hurricane Season of 1992. *Mon. Wea. Rev.*, 122, 517–538.

Pasch, R. J., and E. N. Rappaport, 1995. Atlantic Hurricane Season of 1993. *Mon. Wea. Rev.*, 123, 871–886.

Avila, L. A., and E. N. Rappaport, 1996. Atlantic Hurricane Season of 1994. *Mon. Wea. Rev.*, 124, 1558–1578.

Lawrence, M. B., B. M. Mayfield, L. A. Avila, R. J. Pasch, and E. N. Rappaport, 1998. Atlantic Hurricane Season of 1995. *Mon. Wea. Rev.*, 126, 1124–1151.

Pasch, R. J., and L. A. Avila, 1999. Atlantic Hurricane Season of 1996. *Mon. Wea. Rev.*, 127, 581–610.

Rappaport, E. N., 1999. Atlantic Hurricane Season of 1997. *Mon. Wea. Rev.*, 127, 2012–2026.

Joint Typhoon Warning Center's Annual Tropical Cyclone Summary

The Joint Typhoon Warning Center (JTWC) provides tropical cyclone forecasting support to the U.S. military and to allies within the Pacific and Indian Ocean basins under the auspices of the Naval Pacific Meteorology and Oceanography Center in Hawaii. JTWC was founded May 1, 1959, when the U.S. commander-in-chief of the Pacific forces directed that a single tropical cyclone warning center be established for the western North Pacific region. This region stretches from 180 degrees E longitude westward across the Western Pacific and Indian Oceans in the Northern Hemisphere, and from the oceans near Australia westward to the eastern coast of Africa in the Southern Hemisphere.

Every year, the JTWC staff prepares an annual report summarizing hurricane activity in the region. Included are storm tracks, a history of each storm, and satellite pictures. Tables and graphs present JTWC's intensity and track forecasts, as well as forecast errors. Ongoing research activities are summarized.

Summaries are written for each year starting in 1959 and extending through 1995. These books can be read at some university libraries; U.S. government libraries, such as the Hurricane Research Division and Naval Research Lab; forecast centers worldwide, such as the Japan Meteorological Agency; the Naval Postgraduate School; and at JTWC. Starting in 1996, JTWC stopped printing the annual issue and made them available on their homepage at http://www.npmoc.navy.mil. An example reference for 1994 is shown below (the other references are the same format, just a different number of pages for each year).

Staff of the Joint Typhoon Warning Center. *1994 Annual Tropical Cyclone Report.* NAVPACMETOCCENWEST/JTWC, Guam, Mariana Islands, 1995. 337 pp.

Annual Summary of Australian Cyclones

The *Australian Meteorological Magazine* has a thorough annual summary of hurricanes that have occurred in the Southeast Indian/Australia and the Australia/Southwest Pacific basins.

Annual Summary of Cyclones in the North Indian Ocean

The Indian journal *Mausam* carries an annual summary of hurricane activity over the North Indian Ocean.

Annual Summary of Tropical Cyclones for All Ocean Basins

The *Mariner's Weather Log* has annual summaries of hurricane activity for all of the global basins. These articles are descriptive and nontechnical.

Natural Disaster Survey Reports

After every major natural disaster, the National Oceanographic and Atmospheric Administration (NOAA) sends a survey team to determine how effectively its warning and detection system performed and to identify systematic strengths and weaknesses so that any necessary improvements can be developed and implemented. These *Natural Disaster Survey Reports* contain useful data, figures, and recommendations not easily found in other publications. A sample of these reports is listed below. Natural Disaster Survey Reports have been prepared for any recent major hurricane affecting the United States, so please contact NOAA for more details. In addition, occasionally a "Committee on Natural Disasters" will conduct on-site studies and prepare reports reflecting their findings, as well as make recommendations on the mitigation of natural disasters. Example references of these *Natural Disaster Studies* for Hurricanes Hugo and Elena are listed below.

Natural Disaster Survey Report. *Hurricane Frederic, August 29– September 13, 1979.* Silver Spring, MD: U.S. Department of Commerce, National Oceanic and Atmospheric Administration, National Weather Service, 1979. 27 pp.

Natural Disaster Survey Report. *Hurricane Gilbert, September 3–16, 1988.* Silver Spring, MD: U.S. Department of Commerce, National Oceanic and Atmospheric Administration, National Weather Service, 1998. 35 pp.

Natural Disaster Survey Report. *Hurricane Andrew, August 23–26, 1992.* Silver Spring, MD: U.S. Department of Commerce, National Oceanic and Atmospheric Administration, National Weather Service, 1993. 131 pp.

Natural Disaster Survey Report. *Hurricane Iniki, September 6–13, 1992.* Silver Spring, MD: U.S. Department of Commerce, National Oceanic and Atmospheric Administration, National Weather Service, 1993. 54 pp.

Natural Disaster Survey Report. *Tropical Storm Alberto, July, 1994.* Silver Spring, MD: U.S. Department of Commerce, National Oceanic and Atmospheric Administration, National Weather Service, 1995. 47 pp.

National Research Council. *Natural Disaster Studies—Hurricane Elena, August 29–September 2, 1985.* Washington, DC: National Academy Press, 1991. 121 pp. ISBN 0-309-04434-0.

National Research Council. *Natural Disaster Studies—Hurricane Hugo. Puerto Rico, the Virgin Islands, and Charleston, South Carolina, September 17–22, 1989.* Washington, DC: National Academy Press, 1994. 276 pp. ISBN 0-309-04475-8.

Storm Data

Storm Data is a list of severe weather observations across the United States and is published on a monthly basis by the National Climatic Data Center. The reports are generated through 1) National Weather Service (NWS) phone calls to damaged regions; 2) reports volunteered to the NWS from emergency management groups (such as the Army Corps of Engineers, the Federal Emergency Management Agency, and the U.S. Geological Survey), law enforcement agencies, the general public, and other credible organizations; and 3) information provided to NWS offices by newspaper clippings. Caution is advised when using *Storm Data,* since it is difficult to verify the accuracy of some reports (i.e., general public reports and newspaper clippings).

With regard to hurricanes, detailed documentation of meteorological data, storm surge heights, general statistical information, and storm history can be found when such a storm makes landfall. To find out information about a particular hurricane, one needs to know the year and month when the hurricane impacted the United States, then look up that particular monthly version of *Storm Data.* National Weather Service offices contain these books, as do some universities and libraries. The books may also be ordered from the National Climatic Data Center.

Project **STORMFURY**

Detailed annual reports were published by NOAA relating to Project STORMFURY from 1962 to 1971. Although it is doubtful extra copies are available on Project STORMFURY anymore, they can be obtained from the NOAA library and at the Hurricane Research Division's library. Their references are:

Project STORMFURY Annual Report 1963. Miami: Department of Commerce, Weather Bureau, and U.S. Department of the Navy, Naval Weather Service, 1964.

Project STORMFURY Annual Report 1964. Miami: Department of Commerce, ESSA, and U.S. Department of the Navy, Naval Weather Service, 1965.

Project STORMFURY Annual Report 1965. Miami: Department of Commerce, ESSA, and U.S. Department of the Navy, Naval Weather Service, 1966.

Project STORMFURY Annual Report 1966. Miami: Department of Commerce, ESSA, and U.S. Department of the Navy, Naval Weather Service, 1967.

Project STORMFURY Annual Report 1967. Miami: Department of Commerce, ESSA, and U.S. Department of the Navy, Naval Weather Service Command, 1968.

Project STORMFURY Annual Report 1968. Miami: Department of Commerce, ESSA, and U.S. Department of the Navy, Naval Weather Service Command, 1969.

Project STORMFURY Annual Report 1969. Miami: Department of Commerce, ESSA, and U.S. Department of the Navy, Naval Weather Service Command, 1970.

Project STORMFURY Annual Report 1970. Miami: Department of Commerce, NOAA, and U.S. Department of the Navy, Naval Weather Service Command, 1971.

Project STORMFURY Annual Report 1971. Miami: Department of Commerce, NOAA, and U.S. Department of the Navy, Naval Weather Service Command, 1972.

How to Order
U.S. Government Technical Reports

Almost all recent (post-1990) U.S. government publications and technical reports can be ordered through the National Technical Information Service (NTIS). For example, one could order many of the technical reports from the National Hurricane Center from NTIS. NTIS is located at 5285 Port Royal Road, Springfield, VA 22161. To order a document, call the NTIS sales desk at 1-800-553-6847. Alternatively, order by fax at (703) 605-6900, or by e-mail at orders@ntis.fedworld.gov. For additional information, visit their website at http://www.ntis.gov.

Electronic and Internet Resources

7

S everal good videos have been produced about hurricanes in the past decade, augmented by home video footage taken during and after hurricane landfall displaying the awesome force of these storms. The hurricane videos that are available are summarized in this chapter. Also listed are several worthwhile videos that discuss other weather phenomena in addition to hurricanes, such as tornadoes and thunderstorms, as well as introductory videos about meteorology. Meteorology CD-ROMS for instructional use are listed, as are hurricane and meteorology websites.

Videos Strictly about Hurricanes

Hurricane!
Length: 1 hr.
Cost: $19.95 plus shipping and handling
Date: 1989
Source: WGBH Boston Video
P.O. Box 2284
South Burlington, VT 05407-2284
Phone: 1-800-255-9424
Fax: (802) 864-9846

http://www.pbs.org/wgbh/nova/novastore.html

Also available from (at a cost of $29.95 plus shipping and handling):

Source: Teacher's Video Company
P.O. Box SCE-4455
Scottsdale, AZ 85261
Phone: 1-800-262-8837
Fax: (602) 860-8650

The video *or* transcript may also be ordered from (price un-known):

Source: 800 All News
Suite 390
1535 Grant Street
Denver, CO 80203
Phone: 1-800-ALL-NEWS or (303) 831-9000
http://www.800-all-news.com

One may also purchase this video as part of a three-video pack-age called "Nature's Fury," which not only includes the hurricane footage but other natural phenomena, such as lightning, torna-does, and earthquakes. This package, at a cost of $49.95 plus ship-ping and handling, can be purchased from:

Source: ShopPBS
6902 Hawthorn Park Drive
Indianapolis, IN 46220
Phone: 1-800-645-4PBS
Fax: (703) 739-8131
E-mail: shop@pbs.org
http://www.pbs.org

The Public Broadcasting Service airs a weekly science show called NOVA. On November 7, 1989, NOVA aired this show, which con-tains excellent Hurricane Camille footage as well as interviews of people who were in this storm. A general description of hurri-canes is included in this program. NOVA follows hurricane fore-casters and researchers as they monitor Hurricane Gilbert, which is the strongest Atlantic storm on record. NOVA shows how NHC forecasters make their forecasts and tags along with the Hurricane Research Division as they fly into Hurricane Gilbert.

The Hurricane of '38
Length: 1 hr.
Cost: Currently not available for purchase
Source: WGBH
125 Western Avenue
Boston, MA 02134
Phone: (617) 492-2777
Fax: (617) 787-0714
E-mail: feedback@wgbh.org
http://www.wgbh.org

The Public Broadcasting Service airs a weekly historical documentary series called "The American Experience," which included this documentary of the New England Hurricane (1938). Rhode Island fishermen, residents, and vacationers recount their experience with this hurricane. Footage of the hurricane during landfall is also included. Currently it is not available on video, but interested readers are encouraged to contact PBS for information on how they can view this series.

**Hurricane Force: A Coastal Perspective and
Anatomy of a Hurricane**
Length: 1 hr.
Cost: $19.95 plus shipping and handling
Date: 1989
Source: University of California Extension
Center for Media and Independent Learning
2000 Center Street, Fourth Floor
Berkeley, CA 94704
Phone: (510) 642-0460
Fax: (510) 643-9271
http://www.cmil.unex.berkeley.edu/media/

This video contains two programs. The first is called *Hurricane Force,* a 29-minute documentary that illustrates the environmental and economic consequences of three recent major hurricanes: Andrew (on coastal Louisiana in 1992), Hugo (on coral reefs in Puerto Rico in 1989), and Iniki (on Kauai, Hawaii, in 1992). Computer animation is incorporated with footage taken by "hurricane chasers" that demonstrates how hurricanes form, progress, and affect coastlines as they make landfall. A second 5-minute program follows called *Anatomy of a Hurricane,* which combines hurricane footage computer animation and graphics to provide a more detailed account of how hurricanes form and develop.

Hurricane Iniki: Through the Eyes of Kauai's People
Length: 56 min.
Cost: $25.00 plus shipping and handling
Date: 1992
Source: 3 Friends in Kona
MSC-121
75-1027 Henry #111A
Kailua-Kona, HI 96740-3137
Phone: 1-800-294-2648
Fax: (808) 328-8018
http://www.aloha.net/~chee/video.htm

Also available from (at a cost of $25.00 plus shipping and handling):

Historical Publications
P.O. Box 22617
Charleston, SC 29413
Phone: 1-800-433-7396
Fax: (803) 849-6717
http://www.weathercatalog.com

This powerful video contains some of the most spectacular footage of hurricane damage because, unlike other storms, such as Andrew and Hugo, Iniki made landfall during daylight hours. This compilation of numerous videos taken by tourists and local media is presented in chronological fashion, showing before, during, and after landfall.

Hurricane Georges
Length: 60 min.
Cost: $25.00 plus shipping and handling
Date: 1999
Source: Historical Publications
P.O. Box 22617
Charleston, SC 29413
Phone: 1-800-433-7396
Fax: (803) 849-6717
http://www.weathercatalog.com

This video shows the impact of Hurricane Georges (1998) on the Florida Keys as documented through the *Key West Citizen* and

Free Press reporters who live on the Keys. Footage of the damage along "houseboat row" and interviews with Key West residents are included.

Hurricane Camille 1969
Length: 60 min.
Cost: $25.00 plus shipping and handling
Date: 1998
Source: Historical Publications
P.O. Box 22617
Charleston, SC 29413
Phone: 1-800-433-7396
Fax: (803) 849-6717
http://www.weathercatalog.com

Originally filmed on 16-millimeter film, this video shows Hurricane Camille's (1969) incredible destruction on the Louisiana, Mississippi, Alabama, and Florida Gulf coasts. This storm was one of only two U.S. hurricanes that made landfall as Category 5 storms.

Hurricane Opal
Length: 60 min.
Cost: $25.00 plus shipping and handling
Date: 1993
Source: Historical Publications
P.O. Box 22617
Charleston, SC 29413
Phone: 1-800-433-7396
Fax: (803) 849-6717
http://www.weathercatalog.com

Video of Hurricane Opal's (1995) impact on the Florida panhandle is shown. Footage includes beach dunes being leveled and Highway 98 being washed away by a 15-foot storm surge.

Deadly Water—Tropical Storm Alberto
Length: 60 min.
Cost: $25.00 plus shipping and handling
Date: 1995
Source: Historical Publications
P.O. Box 22617
Charleston, SC 29413

Phone: 1-800-433-7396
Fax: (803) 849-6717
http://www.weathercatalog.com

This video shows the impact of Tropical Storm Alberto (1994). Alberto produced record-breaking rainfall and floods that took 28 lives in Georgia and 2 in Alabama. This video contains dramatic footage of the flooded regions from Macon to Bainbridge.

Hurricane Erin
Length: 60 min.
Cost: $25.00 plus shipping and handling
Date: 1996
Source: Historical Publications
            P.O. Box 22617
            Charleston, SC 29413
            Phone: 1-800-433-7396
            Fax: (803) 849-6717
            http://www.weathercatalog.com

This video shows the structural damage of Hurricane Erin (1995) along the Pensacola Beach and Navarre Beach in the Florida panhandle. Ironically, the same region would be hit by Hurricane Opal two months later.

Raging Planet: Hurricane
Length: 30 min.
Cost: $29.95 plus shipping and handling
Date: 1996
Source: The Discovery Channel (School Store Section)
            Phone: 1-888-892-3484
            http://school.discovery.com

A general description of hurricanes is included in this program. Footage and a historical account of Hurricane Andrew and such devastating storms as the 1970 Bangladesh storm that killed 300,000 people is included. A documentary of North Carolina residents and emergency preparedness officials is presented as they prepare for Hurricane Bertha and Hurricane Fran in 1996. Footage of the Hurricane Hunters as they fly into these storms is shown.

Savage Skies: Monster of the Deep
Length: 1 hr.

Cost: $29.95 plus shipping and handling
Date: 1992
Source: Teacher's Video Company
P.O. Box SCE-4455
Scottsdale, AZ 85261
Phone: 1-800-262-8837
Fax: (602) 860-8650

This video chronicles the events before, during, and after the landfall of Hurricane Andrew on Miami in 1992. Compelling interviews are conducted with National Hurricane Center forecasters and several people who experienced the wrath of the storm as it destroyed their homes.

Hurricane Andrew
Length: 60 min.
Cost: $25.00 plus shipping and handling
Date: 1993
Source: Historical Publications
P.O. Box 22617
Charleston, SC 29413
Phone: 1-800-433-7396
Fax: (803) 849-6717
http://www.weathercatalog.com

This video displays the horrible damage caused by this century's costliest hurricane.

Hurricane Hugo
Length: 60 min.
Cost: $25.00 plus shipping and handling
Date: 1990
Source: Historical Publications
P.O. Box 22617
Charleston, SC 29413
Phone: 1-800-433-7396
Fax: (803) 849-6717
http://www.weathercatalog.com

This video tracks Hurricane Hugo's (1989) destructive path through San Juan, Puerto Rico before making landfall on the North Carolina coast. Hugo was the second costliest hurricane in U.S. history.

Hurricanes of the 80's
Length: 80 min.
Cost: $25.00 plus shipping and handling
Date: 1990
Source: Historical Publications
P.O. Box 22617
Charleston, SC 29413
Phone: 1-800-433-7396
Fax: (803) 849-6717
http://www.weathercatalog.com

This video shows footage of hurricanes Elena (1985), Gilbert (1988), Diana (1984), and Hugo (1989).

Jason and Robin's Awesome Hurricane Adventure Companion Video
Length: 10 min.
Cost: $5.00 plus shipping and handling
Source: American Red Cross
P.O. Box 37243
Washington, DC 20013
Phone: 1-800-435-7669
http://www.redcross.org

This is the companion video to the 12-page, four-color workbook called *Jason and Robin's Awesome Hurricane Adventure* for children in third to sixth grade. It covers hurricane facts, hazard avoidance, planning, supplies, and what to do when a hurricane watch or warning is issued.

Against the Wind: Protecting Your Home from Hurricane Wind Damage
Length: 18 min.
Cost: $5.00 plus shipping and handling
Source: American Red Cross
P.O. Box 37243
Washington, DC 20013
Phone: 1-800-435-7669
http://www.redcross.org

This video demonstrates how to inspect and make simple changes within homes to mitigate the potentially devastating effects of hurricane wind damage.

Home Preparedness for Hurricanes
Length: 6 min.
Cost: $5.00 plus shipping and handling
Source: American Red Cross
P.O. Box 37243
Washington, DC 20013
Phone: 1-800-435-7669
http://www.redcross.org

This video, intended for general audiences, discusses how to prepare ahead of time for a hurricane.

Before the Wind Blows
Length: 12 min.
Cost: $5.00 plus shipping and handling
Source: American Red Cross
P.O. Box 37243
Washington, DC 20013
Phone: 1-800-435-7669
http://www.redcross.org

This video, intended for adults, discusses hurricane information, planning decisions, evacuation information, and how the Red Cross responds to hurricanes.

Hurricane Information Guide for Coastal Residents
Length: 17 min.
Cost: $4.40 plus shipping and handling
Source: American Red Cross
P.O. Box 37243
Washington, DC 20013
Phone: 1-800-435-7669
http://www.redcross.org

This video, featuring Maury Povich, gives information on how to prepare one's home and family for a hurricane.

General Weather Videos That Include Information about Hurricanes

Eyewitness: Natural Disasters
Length: 35 min.

Cost: $14.98 plus shipping and handling
Date: 1989
Source: ShopPBS
6902 Hawthorn Park Drive
Indianapolis, IN 46220
Phone: 1-800-645-4PBS
Fax: (703) 739-8131
E-mail: *shop@pbs.org*
http://www.pbs.org

This video displays most of the natural forces, from hurricanes to volcanoes. Rare footage shows violent natural phenomena, such as tornadoes and avalanches.

Killer Weather

Length: 30 min.
Cost: $139.00
Date: 1997
Source: Insight Media
2162 Broadway
New York, NY 10024-6620
Phone: 1-800-233-9910 or (212) 721-6316
http://www.insight-media.com

This program features footage of the most dramatic aspects of weather. It examines such phenomena as hurricanes, tornadoes, hailstorms, and floods and investigates extreme conditions of heat, dryness, sun, and snow.

Meteorology: Trying to Bat a Thousand

Length: 20 min.
Cost: $195.00
Date: 1994
Source: Insight Media
2162 Broadway
New York, NY 10024-6620
Phone: 1-800-233-9910 or (212) 721-6316
http://www.insight-media.com

This video examines how meteorologists use information from Doppler radar, weather balloons, remote data collection sites, satellites, commercial aircraft, and computers to predict weather. It discusses tornadoes, hurricanes, and thunderstorms.

Weather: The Chaos Which Surrounds Us
Length: 25 min.
Cost: $135.00
Date: 1996
Source: Insight Media
2162 Broadway
New York, NY 10024-6620
Phone: 1-800-233-9910 or (212) 721-6316
http://www.insight-media.com

This program describes the general forces behind weather. It introduces the viewer to the flow of energy in the atmosphere that leads to weather patterns and details how much violent weather phenomena, such as thunderstorms, lightning, tornadoes, and hurricanes, develop.

Wonders of Weather
Length: 60 min.
Cost: $34.95 plus shipping and handling
Source: The Discovery Channel (School Store Section)
Phone: 1-888-892-3484
http://school.discovery.com

This video shows dramatic footage of nature at its most fierce, including tornadoes and hurricanes. How meteorologists use the latest technology to monitor the weather is shown. Footage of storm chasers as they pursue a tornado in action is presented. Learn how the oceans influence weather and how the collision of such factors as temperature and barometric pressure can produce spectacular and frightening phenomena.

General Weather CD-ROMs for Instructional Use

Operation Weather Disaster
Cost: $39.95 plus shipping and handling
Source: The Discovery Channel (School Store Section)
Phone: 1-888-892-3484
http://school.discovery.com

Students from sixth to twelfth grade can play in teams or individually as their knowledge of weather and problem-solving skills move them through seven levels. A teacher's guide gives classroom activities to accompany each learning level. A five-CD pack is also available so that multiple groups of students can play at once.

The World's Weather

Cost: $89.00 plus shipping and handling
Source: Films for the Humanities & Sciences
P.O. Box 2053
Princeton, NJ 08543-2053
Phone: 1-800-257-5126
Fax: (609) 275-3767
E-mail: custserv@films.com
http://www.films.com

This CD-ROM provides a comprehensive look at weather and climate. Using examples from all over the globe, it examines how different kinds of weather affect the lives of people and other organisms. The program is divided into four sections: Processes in the Atmosphere, Changing Seasons, Weather and Climate, and Life in Different Climate Zones. Each section incorporates photographs, drawings, video, and animation sequences coupled with an easy-to-use interface. Text and images can be extracted for student report making.

Weather Tracker's Kit

Cost: $89.00 plus shipping and handling
Source: Weather Affects
440 Middlesex Road
Tyngsboro, MA 01879
Phone: 1-800-317-3666
Fax: (978) 649-8387
http://wxstore.weather.com/wxstore/index.html
(under "Kid's Corner" section)

The Weather Tracker's kit comes complete with everything young weather enthusiasts need to pursue their hobby, as well as a CD-ROM describing the basic science behind the weather station. Included is a five-function weather station that measures wind speed and direction, rainfall, temperature, and wind chill. The accompanying 80-page Weather Tracker's Handbook and cloud chart guides the student through cloud identification, lightning safety, forecasting, and so on. This product is suitable for students aged 8 years and older.

Internet/Web Sources

The American Meteorological Society
http://www.ametsoc.org/AMS

The American Meteorological Society (AMS) is a nonprofit, professional society for meteorologists and people in related fields. Those interested in learning about meteorology as a career may find the section titled "Challenges for our Changing Atmosphere" interesting. Information about the AMS educational program for elementary and secondary schools, called "Project AT-MOSPHERE," is provided. Several monographs on hurricanes are available for purchase here.

The American Red Cross
http://www.redcross.org

The American Red Cross is a humanitarian organization, led by volunteers, that provides relief to victims of disasters and helps prevent, prepare for, and respond to emergencies. This site contains information on hurricane preparedness and hurricane relief in disaster areas. Many articles can be downloaded in pdf format.

Gray, William M. **Atlantic Seasonal Hurricane Activity Forecasts**
http://tropical.atmos.colostate.edu

Every December, April, June, and August, Dr. Bill Gray and colleagues issue forecasts on Atlantic hurricane activity. These highly anticipated statements discuss in detail the forecast methodology and how different forecast parameters (such as El Niño and African rainfall) are expected to affect the upcoming hurricane season. Archives of previous forecasts as well as verifications are also available.

The Cooperative Institute for Meteorological Satellite Studies Tropical Cyclone Homepage
http://cimss.ssec.wisc.edu/tropic

This site displays state-of-the-art satellite imagery of hurricanes and the tropics around the world, with a research focus on winds derived from cloud motions. Archived hurricane imagery is also available on this site. NHC and JTWC forecasts are also posted on this site.

The Federal Emergency Management Agency
http://www.fema.gov

The Federal Emergency Management Agency (FEMA) is an independent agency of the federal government. Its mission is to reduce life and property losses through mitigation and preparedness programs. FEMA is also called in to help when the president declares a region a disaster area, such as coastal regions impacted by a hurricane. This site provides extensive documentation on hurricane preparedness and hurricane recovery issues. Current weather statements are also available form the National Hurricane Center.

Landsea, Chris W. **Frequently Asked Questions: Hurricanes, Typhoons, and Tropical Cyclones**
http://www.aoml.noaa.gov/hrd/tcfaq

This up-to-date, informative website contains hurricane definitions, answers to specific questions regarding hurricanes, statistics about tropical cyclones in all ocean basins, a listing of websites pertaining to tropical cyclones, and references for more information. It is a popular and highly recommended website. One may also read about the Hurricane Research Division (http://www.aoml.noaa.gov/hrd) on this site.

The Hurricane Hunters
http://www.hurricanehunters.com

The 53rd Weather Reconnaissance Squadron, also known as the Hurricane Hunters, flies into Atlantic tropical disturbances, tropical depressions, tropical storms, and hurricanes on a routine basis and relays this information to the National Hurricane Center. This website includes aircraft reports, pictures of flights inside hurricanes, informative information about the Hurricane Hunters, and educational links.

Intellicast
http://www.intellicast.com

Intellicast is a product of the Weather Service International and provides extensive specialized weather information to help plan all outdoor and weather sensitive activities. This site provides excellent satellite and radar images. A tropical weather section is also included.

Joint Typhoon Warning Center
http://www.npmocw.navy.mil

This site contains the latest JTWC forecast statements. One may also read their *Annual Tropical Cyclone Report* (beginning in 1996), or download tropical cyclone historical datasets containing positions and intensities for West Pacific and Australian storms.

The National Data Buoy Center
http://www.ndbc.noaa.gov

This is an excellent source for obtaining real-time surface weather information over the ocean. It is also interesting to look at buoys when a hurricane is nearby to obtain wind and pressure information (although some buoys do not survive hurricane conditions and stop transmitting data). This site also provides archived buoy data.

The National Hurricane Center
http://www.nhc.noaa.gov

This site contains the latest NHC forecast statements, as well as satellite imagery, reconnaissance reports, historical data, educational material, general information about the forecast procedures, and a description of the forecast facilities.

National Oceanographic and Atmospheric Administration
http://www.noaa.gov

This site is a good starting point for obtaining exclusive weather facts, educational material, weather observations, and forecasts. It contains many links to other NOAA branches, such as the National Weather Service (http://www.nws.noaa.gov).

The Naval Research Laboratory Tropical Cyclone Homepage
http://www.nrlmry.navy.mil/sat-bin/tc_home

This site displays state-of-the-art satellite imagery of hurricanes and the tropics around the world, including scatterometer winds, winds derived from cloud motions, and estimated rainfall rates.

The 1900 Storm—Remembering the Galveston Hurricane, September 8–9, 1900
http://www.1900storm.com

This site is a service of the *Galveston County Daily News,* in conjunction with the Galveston 1900 Storm Commemoration Committee. This website was developed in preparation for the commemoration of the 100th anniversary of the 1900 hurricane and the celebration of the rebuilding of Galveston afterwards. This site contains many pictures of Galveston after the hurricane and vintage film footage of the storm's aftermath taken by one of Thomas Edison's assistants.

NOAA Aircraft Operations Center
http://www.nc.noaa.gov/aoc.html

The Aircraft Operations Center (AOC) provides the aircraft equipped with the scientific instruments required for hurricane research flights by the Hurricane Research Division. This site describes in detail the aircraft specifications. This site also discusses the NOAA Corps, the smallest of the seven uniformed services of the United States. The AOC pilots are members of the NOAA Corps.

Unisys Weather
http://weather.unisys.com

This weather site from Unisys Corporation provides a complete source of graphical weather information. It is intended to satisfy the needs of the weather professional but can be a tool for the casual user as well. The graphics and data are displayed as a meteorologist would expect to see it. For novice users, there are detailed explanation pages to guide them through the various plots, charts, and images. This site also includes a thorough archive of Atlantic, East Pacific, and West Pacific hurricane track graphics, historical track and intensity data, and some satellite imagery of well-known Atlantic storms.

The Weather Channel
http://www.weather.com

This cable media outlet provides excellent, understandable weather information and graphics on its website. In addition, meteorologists provide updated reports on tropical activity. Hurricane graphics and satellite imagery of the topics are shown, and educational hurricane information is available. Short educational videos may also be downloaded.

The Weather Underground
http://www.wunderground.com

An excellent source of quick, accessible National Weather Service observations and forecasts, the Weather Underground also provides the latest information from the National Hurricane Center, the Central Pacific Hurricane Center, and the Joint Typhoon Warning Center. A variety of items may also be purchased from its "weather store." A similar site is http://blueskies.sprl.umich.edu, which is the University of Michigan Weather Underground. The latter site also contains a list of over 300 other weather-related Internet websites and another list of vendors who sell weather software. Two educational programs for teachers, "Kids as Global Scientists" and "Blue-skies," are also there.

Acronyms and Abbreviations

Acronyms

AMS	American Meteorological Society
AOC	NOAA Aircraft Operations Center
CIMSS	Cooperative Institute for Meteorological Satellite Studies
FEMA	Federal Emergency Management Agency
GFDL	Geophysical Fluid Dynamics Laboratory
HRD	Hurricane Research Division
JTWC	Joint Typhoon Warning Center
McIDAS	Man computer Interactive Data Access System
NASA	National Aeronautics and Space Administration
NCDC	National Climatic Data Center
NCEP	National Centers for Environmental Prediction
NESDIS	National Environmental Satellite, Data and Information Service
NHC	National Hurricane Center
NOAA	National Oceanic and Atmospheric Administration
NWA	National Weather Association
NWS	National Weather Service
QBO	Quasi-Biennial Oscillation
WMO	World Meteorological Organization

Abbreviations for
Meteorological Journals

Bull. Amer. Meteor. Soc.	*Bulletin of the American Meteorological Society*
Geo. Res. Letters	*Geophysical Research Letters*
J. Climate	*Journal of Climate*
J. Atmos. Sci.	*Journal of the Atmospheric Sciences*
Mar. Wea. Sci.	*Mariner's Weather Log*
Mon. Wea. Rev.	*Monthly Weather Review*
Nature	*Nature*
Wea. Forecasting	*Weather and Forecasting*
Weatherwise	*Weatherwise*

Conversion Tables

Adapted from Holland (1993), Ahrens (1994), and author's class notes.

Length

1 kilometer (km) = 1000 m
      = 3281 ft
      = 0.621 mi
      = 0.54 nm
1 statute mile (mi) = 5280 ft
      = 1609 m
      = 1.609 km
      = 0.87 nm
1 nautical mile (nm) = 6080 ft
      = 1853 m
      = 1.853 km
      = 1.15 mi
1 meter (m) = 100 cm
      = 3.28 ft
      = 39.37 in
1 foot (ft) = 12 in
      = 30.48 cm
      = 0.305 m
1 centimeter (cm) = 0.39 in
      = 0.01 m
      = 10 mm

1 millimeter (mm) = 0.1 cm
    = 0.001 m
    = 0.039 in
1 inch (in) = 2.54 cm
    = 0.08 ft

Approximate conversion between latitude, longitude, and distance (use with some caution)

1 degree latitude ≈ 111.137 km
    ≈ 60 nm
    ≈ 69 mi
1 degree longitude ≈ 111.137 km × cosine(latitude)
    ≈ 60 nm × cosine(latitude)
    ≈ 69 mi × cosine(latitude)

More precise distance calculations between two latitude and two longitude points

Let point 1 be latitude ϕ_1 and longitude θ_1, and let point 2 be latitude ϕ_2 and longitude θ_2. To accurately compute the distance between these two locations, one computes the distance along a "great circle" using the following steps:

Step 1: Compute C

$$C = \sin\phi_1 \times \sin\phi_2 + \cos\phi_1 \times \cos\phi_2 \times \cos(\theta_2 - \theta_1)$$

Step 2: Compute A (make sure the quantity A is in radians between 0 and 3.14)

$$A = \cos^{-1}C$$

Step 3: Compute distance

    distance=6378 km × A
    distance=3444 nm × A
    distance=3963 mi × A

These computations assume the earth is a perfect sphere, which introduces a small error of about 1 percent.

Area

1 square centimeter (cm^2) = 0.15 in^2
1 square inch (in^2) = 6.45 cm^2
1 square meter (m^2) = 10.76 ft^2
1 square foot (ft^2) = 0.093 m^2

Volume

1 cubic centimeter (cm^3) = 0.06 in^3
1 cubic inch (in^3) = 16.39 cm^3
1 liter (l) = 1000 cm^3
    = 0.264 gallons

Time

1 minute (min) = 60 s
1 hour (hr) = 60 min
    = 3600 s
1 day (d) = 24 hr
    = 1440 min
    = 86400 s

Speed

1 knot (kt) = 1 nautical mi/hr

= 1.15 statute mi/hr, also written as 1.15 mph
= 0.513 m/s
= 1.85 km/hr
1 mile per hour (mi/hr, or mph) = 0.87 kt
= 0.45 m/s
= 1.61 km/hr
1 kilometer per hour (km/hr) = 0.54 kt
= 0.62 mi/hr, or 0.62 mph
= 0.28 m/s
1 meter per second (m/s) = 1.94 kt
= 2.24 mi/hr, or 2.24 mph
= 3.60 km/hr

Mass

1 gram (g) = 0.035 ounce
= 0.0022 lb
1 kilogram (kg) = 1000 g
= 2.2 lb

Pressure

1 millibar (mb) = 1000 dynes/cm^2

= 0.75 millimeters of mercury (mm Hg)
= 0.02953 inch of mercury (in Hg)
= 0.01450 pound per square inch (lb/in^2)
= 100 Pascals (Pa)
1 inch of mercury = 33.865 mb
1 millimeter of mercury = 1.3332 mb
1 Pascal = 0.01 mb
1 hectopascal (hPa) = 1 mb
1 kilopascal (kPa) = 10 mb
1 standard atmosphere = 1013.25 mb
= 760 mm Hg
= 29.92 in Hg
= 14.7 lb/in^2

Conversion °C to °F

$F = {}^9/_5C + 32$
$C = {}^5/_9(F - 32)$

Conversion °K to °C

$C = K - 273.15$
$K = C + 273.15$

References

Ahrens, C. D., 1994. *Meteorology Today.* West Publishing Co., 591 pp.

Holland, G. J., 1993. *Global Guide to Tropical Cyclone Forecasting.* World Meteorological Organization Technical Document, WMO/TD No. 560, Tropical Cyclone Programme, Report No. TCP-31, Geneva, Switzerland. Also available at http://www.bom.gov.au/bmrc (look for additional links under the mesoscale section to find this book).

The Modified Beaufort Scale

I n 1831, Rear Admiral Sir Francis Beaufort developed a scale numbering from 0 to 12 based on the sea's impact on a ship (Kinsman 1990, 1991). From this, a common name for wind strength was assigned (e.g., "gentle breeze," "fresh gale," etc.). Since then, the scale has been modified several times to estimate wind speed by land or sea observations. In addition, one may also use the table to determine wave height offshore for a given wind speed. The scale below is adapted from Holland (1993) and the official *National Weather Service Mariners Observing Handbook; No. 1—Marine Surface Weather Observations.* Wave heights should be used for rough estimates only, because wave height depends on the duration of the wind speed, how large an area the wind is affecting, and whether the waves have been generated locally or some distance from the windy region. Also, the values given are average wave heights, and sporadic higher waves are possible. Caution should also be shown using the Beaufort scale outside U.S. waters, because the United States uses 1-minute average winds and other countries use 10-minute averages; using 10-minute averages will result in smaller wind speeds.

The Modified Beaufort Scale

Beaufort Number	Description	Wind Speed Mph	Wind Speed Knots	Km/hr	Mean Wind Pressure (lb/ft²)	Wind Effects Observed on Land	Possible Wave Height in Deep Water Offshore in Ft (use for rough estimate only)	Wind Effects Observed at Sea
0	Calm	Less than 1	Less than 1	Less than 2	0	Calm; smoke rises vertically	Less than ½ foot	Sealike mirror
1	Light air	1–3	1–3	2–6	0.01	Direction of wind shown by smoke drift but not by vanes	Less than ½ foot	Slight ripples; no foam crests
2	Light breeze	4–7	4–6	7–11	0.08	Wind felt on face; leaves rustle; vanes moved by wind; flags stir	½ to 1 foot	Small wavelets, still short but more pronounced; crests have a glassy appearance and do not break
3	Gentle breeze	8–12	7–10	12–19	0.28	Leaves, small twigs in constant motion; wind extends light flag	2 to 3	Large wavelets; crests breaking; foam glassy; scattered whitecaps
4	Moderate breeze	13–18	11–16	20–29	0.67	Raises dust, loose paper; small branches move; flags flap	3½ to 5	Small waves become larger; fairly frequent whitecaps
5	Fresh breeze	19–24	17–21	30–39	1.3	Small trees with leaves begin to sway; flags ripple; crested wavelets form on inland waters	6 to 8½	Moderate waves form many whitecaps; chance of some spray
6	Strong breeze	25–31	22–27	40–50	2.3	Large branches in motion; whistling heard in utility wires; umbrellas used with difficulty	9½ to 12	Large waves form; foam crests more extensive; some spray
7	Near gale	32–38	28–33	51–61	3.6	Whole trees in motion; inconvenient walking against wind; flags extend	13½ to 19	Sea heaps up; some foam from waves blows in streaks
8	Gale	39–46	34–40	62–74	5.4	Twigs break off trees; walking is difficult	18 to 25	Moderately high waves of greater length; well-marked streaks of foam
9	Strong gale	47–54	41–47	75–87	7.7	Slight structural damage occurs; signs and antennas blown down	23 to 32	High waves; dense foam streaks; crests of waves begin to topple,

(continues)

The Modified Beaufort Scale
(continued)

Beaufort Number	Description	Mph	Wind Speed Knots	Km/hr	Mean Wind Pressure (lb/ft²)	Wind Effects Observed on Land	Possible Wave Height in Deep Water Offshore in Ft (use for rough estimate only)	Wind Effects Observed at Sea
								tumble, and roll over; spray may affect visibility
10	Storm	55–63	48–55	88–101	10.5	Trees uprooted; considerable structural damage occurs	29 to 41	Very high waves with long overhanging crests; foam blown in white dense streaks; on the whole, the sea takes a white appearance; visibility affected
11	Violent storm	64–73	56–63	102–119	14.0	Widespread damage	37 to 52	Exceptionally high waves (small and medium-sized ships might be lost to view behind the waves); sea completely covered with long white patches of foam; edges of waves blow into froth everywhere; visibility affected
12	Hurricane	74 and over	64 and over	120 and over	Over 17.0	Extensive damage	45 or greater	Air filled with foam and spray; sea completely white with driving spray; visibility very seriously affected

Estimating Wave Height in Shallow Water

The Beaufort table cannot be used to estimate wave heights near the coast in shallow water. As waves enter water less than 65 feet deep, they begin to "feel" the ocean bottom. As waves propagate closer to the coast, eventually they will "break." New waves may form as they propagate inland, but they will be smaller. One can estimate the maximum possible wave height as roughly 75 percent of the water depth (including astronomical tide and storm surge). However, exceptions can occur in regions where the water is deep near the coast. Local topography also plays a key role, as waves may be diffracted or concentrated by points and inlets. In regions where sharp vertical cliffs occur near deep water, waves do not break but "reflect" off the cliff; when combined with the lifting mechanism of a storm surge, the waves can overtop the cliff, causing considerable destruction.

References

Holland, G. J., 1993. In *Global Guide to Tropical Cyclone Forecasting*. World Meteorological Organization Technical Document, WMO/TD No. 560, Tropical Cyclone Programme, Report No. TCP-31, Geneva, Switzerland. Also available at http://www.bom.gov.au/bmrc (look for additional links under the mesoscale section to find this book).

Kinsman, B., 1990. Who Put the Wind Speeds in Admiral Beaufort's Force Scale? Part I—the Original Scale. *Mar. Wea. Log* 34, 2–8.

Kinsman, B., 1991. Who Put the Wind Speeds in Admiral Beaufort's Force Scale? Part II—the New Scale. *Mar. Wea. Log* 35, 12–18.

National Weather Service, 1991. Marine Surface Weather Observations. In *National Weather Service Handbook; No. 1*. Silver Spring, MD: U.S. Department of Commerce, National Oceanic and Atmospheric Administration, National Weather Service, Office of Systems Operations, Observing Systems Branch.

Tracking Hurricanes and Understanding National Hurricane Center Forecasts

Tracking a hurricane is an interesting pastime for many people as they track where a hurricane has been and postulate where it may go. It is also an educational exercise for many students because it integrates geography, mapping skills, and science. When a tropical depression, tropical storm, or hurricane exists in the Atlantic or East Pacific, the National Hurricane Center will issue forecast statements. One may plot the storm's position with a marker on a hurricane tracking chart, which lets one follow the storm's path. One may also plot the storm's projected path based on NHC's forecasts. Hurricane tracking charts may be ordered from the National Weather Service (see Chapter 5 for points of contact). In addition, coastal residents can often find tracking charts at many local stores.

Hurricane center positions are given by latitude (for example, 28.4 degrees North) and longitude (for example, 88.7 degrees West) by the National Hurricane Center in advisories issued every six hours at 5 A.M., 11 A.M., 5 P.M., and 11 P.M. eastern standard time. The current intensity is also given, as

251

well as the predicted track and intensity. As a hurricane nears landfall, intermediate advisories may also be issued in three-hour increments between the forecast advisories.

There are four kinds of statements issued by NHC every six hours: (1) a forecast advisory; (2) a public advisory; (3) discussion; and (4) strike probability. These statements can be viewed at many weather websites on the Internet, as well as the National Hurricane Center website at http://www.nhc.noaa.gov. Examples for Hurricane Georges as it approached the Louisiana and Mississippi coasts are shown below.

Forecast Advisory

Forecast advisories contain detailed information about the hurricane's current location, intensity, and wind distribution. The wind distribution (in nautical miles) is given in "radius of 35 knot" winds and "radius of 50 knot" winds. Mariners generally avoid winds greater than 35 knots, and the U.S. Navy is required to avoid winds of these magnitudes. Forecast advisories also predict these parameters out to three days. When landfall is expected, watches and warnings are also included in the forecast advisory.

HURRICANE GEORGES FORECAST/ADVISORY NUMBER   49
NATIONAL WEATHER SERVICE MIAMI FL   AL0798
1500Z   SUN   SEP 27   1998

A HURRICANE WARNING IS IN EFFECT FROM
MORGAN CITY LOUISIANA TO PANAMA CITY
FLORIDA. A HURRICANE WARNING MEANS THAT
HURRICANE CONDITIONS ARE EXPECTED IN THE
WARNED AREA WITHIN 24 HOURS. PREPARATIONS
TO PROTECT LIFE AND PROPERTY SHOULD BE
RUSHED TO COMPLETION.

A TROPICAL STORM WARNING AND A HURRICANE
WATCH ARE IN EFFECT FROM EAST OF PANAMA CITY
FLORIDA TO ST. MARKS FLORIDA. A HURRICANE
WATCH IS IN EFFECT FROM WEST OF MORGAN CITY
TO INTRACOASTAL CITY LOUISIANA.

HURRICANE CENTER LOCATED NEAR 28.4N 88.0W AT
27/1500Z POSITION ACCURATE WITHIN 25 NM

PRESENT MOVEMENT TOWARD THE NORTHWEST OR
315 DEGREES AT 7 KT

ESTIMATED MINIMUM CENTRAL PRESSURE 963 MB
MAX SUSTAINED WINDS   95 KT WITH GUSTS TO 115 KT
64 KT. 80NE   40SE   20SW   60NW
50 KT. 100NE   90SE   30SW   75NW
34 KT. 130NE   150SE   90SW   100NW
12 FT SEAS..130NE   150SE   90SW   100NW
ALL QUADRANT RADII IN NAUTICAL MILES

REPEAT . . . CENTER LOCATED NEAR 28.4N 88.0W AT
27/1500Z
AT 27/1200Z CENTER WAS LOCATED NEAR 28.2N
87.8W

FORECAST VALID 28/0000Z 28.9N   88.7W
MAX WIND 100 KT . . . GUSTS 120 KT
64 KT . . . 90NE   40SE   20SW   60NW
50 KT . . .100NE   90SE   30SW   75NW
34 KT . . .130NE   150SE   90SW   100NW

FORECAST VALID 28/1200Z 29.6N   89.5W
MAX WIND 100 KT . . . GUSTS 120 KT
64 KT . . . 80NE   40SE   20SW   60NW
50 KT . . . 100NE   90SE   30SW   75NW
34 KT . . . 130NE   150SE   90SW   100NW

FORECAST VALID 29/0000Z 30.3N   89.9W . . . INLAND
MAX WIND 80 KT . . . GUSTS 95 KT
64 KT . . . 80NE   40SE   20SW   60NW
50 KT . . . 100NE   90SE   30SW   75NW
34 KT . . . 130NE   150SE   90SW   100NW

STORM SURGE FLOODING OF 10 TO 15 FEET . . .
LOCALLY UP TO 17 FEET AT THE HEADS OF BAYS . . .
ABOVE NORMAL TIDE LEVELS IS POSSIBLE IN THE
WARNED AREA AND WILL BE ACCOMPANIED BY
LARGE AND DANGEROUS BATTERING WAVES.

SMALL CRAFT FROM INTRACOASTAL CITY LOUISIANA WESTWARD AND SOUTHWARD ALONG THE COAST OF TEXAS SHOULD REMAIN IN PORT. SMALL CRAFT ALONG THE WEST COAST OF THE FLORIDA PENINSULA SHOULD REMAIN IN PORT UNTIL WINDS AND SEAS SUBSIDE.

REQUEST FOR 3 HOURLY SHIP REPORTS WITHIN 300 MILES OF 28.4N 88.0W

EXTENDED OUTLOOK . . . USE FOR GUIDANCE ONLY . . . ERRORS MAY BE LARGE

OUTLOOK VALID 29/1200Z 30.8N 89.9W . . . INLAND MAX WIND 65 KT . . . GUSTS 80 KT
50 KT . . . 50NE 75SE 25SW 25NW

OUTLOOK VALID 30/1200Z 31.5N 89.5W . . . INLAND MAX WIND 50 KT . . . GUSTS 60 KT
50 KT . . . 50NE 75SE 25SW 25NW

NEXT ADVISORY AT 27/2100Z

GUINEY

Public Advisory

A public advisory contains the hurricane's current position and intensity, as well as general commentary concerning watches, warnings, future track and intensity, wind distribution, potential tornadoes, potential storm surge, and potential rainfall when landfall is possible or likely.

BULLETIN
HURRICANE GEORGES ADVISORY NUMBER 49
NATIONAL WEATHER SERVICE MIAMI FL
10 AM CDT SUN SEP 27 1998

. . . DANGEROUS HURRICANE GEORGES CLOSING IN ON THE CENTRAL GULF COAST . . .

A HURRICANE WARNING IS IN EFFECT FROM
MORGAN CITY LOUISIANA TO PANAMA CITY
FLORIDA. A HURRICANE WARNING MEANS THAT
HURRICANE CONDITIONS ARE EXPECTED IN THE
WARNED AREA WITHIN 24 HOURS. PREPARATIONS TO
PROTECT LIFE AND PROPERTY SHOULD BE RUSHED
TO COMPLETION . . . AND ADVICE FROM LOCAL
EMERGENCY MANAGEMENT OFFICIALS SHOULD BE
CLOSELY FOLLOWED.

A TROPICAL STORM WARNING AND A HURRICANE
WATCH ARE IN EFFECT FROM EAST OF PANAMA CITY
FLORIDA TO ST. MARKS FLORIDA. A HURRICANE
WATCH IS IN EFFECT FROM WEST OF MORGAN CITY
TO INTRACOASTAL CITY LOUISIANA.

AT 10 AM CDT . . . 1500Z . . . THE CENTER OF
HURRICANE GEORGES WAS LOCATED NEAR
LATITUDE 28.4 NORTH . . . LONGITUDE 88.0 WEST.
THIS POSITION IS ABOUT 80 MILES SOUTHEAST OF
THE MOUTH OF THE MISSISSIPPI RIVER AND ABOUT
175 MILES SOUTHEAST OF NEW ORLEANS LOUISIANA.

GEORGES IS MOVING TOWARD THE NORTHWEST
NEAR 8 MPH AND THIS GENERAL MOTION IS
EXPECTED TO CONTINUE WITH A GRADUAL
DECREASE IN FORWARD SPEED. THIS WOULD BRING
THE CORE OF THE HURRICANE NEAR THE MOUTH OF
THE MISSISSIPPI RIVER LATER TODAY. DO NOT FOCUS
ON THE PRECISE LOCATION AND TRACK OF THE
CENTER. THE HURRICANES DESTRUCTIVE WINDS . . .
RAIN . . . AND STORM SURGE COVER A WIDE SWATH.

MAXIMUM SUSTAINED WINDS ARE NEAR 110 MPH
WITH HIGHER GUSTS. GEORGES IS STRONG CATEGORY
TWO HURRICANE ON THE SAFFIR-SIMPSON
HURRICANE SCALE AND COULD REACH CATEGORY
THREE STATUS BEFORE LANDFALL.

HURRICANE FORCE WINDS EXTEND OUTWARD UP TO
90 MILES FROM THE CENTER . . . AND TROPICAL
STORM FORCE WINDS EXTEND OUTWARD UP TO 175

MILES. RAINBANDS OF GEORGES ARE SPREADING ACROSS PORTIONS OF THE WARNING AREA AND HURRICANE FORCE WINDS SHOULD BEGIN TO AFFECT THE AREA LATER TODAY.

AN AIR FORCE RESERVE UNIT HURRICANE HUNTER PLANE REPORTED A MINIMUM CENTRAL PRESSURE OF 963 MB . . . 28.44 INCHES.

STORM SURGE FLOODING OF 10 TO 15 FEET ABOVE NORMAL TIDE LEVELS . . . AND UP TO 17 FEET AT THE HEADS OF BAYS . . . IS POSSIBLE IN THE WARNED AREA AND WILL BE ACCOMPANIED BY LARGE AND DANGEROUS BATTERING WAVES.

FLOODING RAINS OF 15 TO 25 INCHES . . . WITH LOCALLY HIGHER AMOUNTS . . . ARE LIKELY IN ASSOCIATION WITH THIS SLOW-MOVING HURRICANE.

ISOLATED TORNADOES ARE POSSIBLE EAST AND NORTHEAST OF THE TRACK OF GEORGES.

SMALL CRAFT FROM INTRACOASTAL CITY LOUISIANA WESTWARD AND SOUTHWARD ALONG THE COAST OF TEXAS SHOULD REMAIN IN PORT. SMALL CRAFT ALONG THE WEST COAST OF THE FLORIDA PENINSULA SHOULD REMAIN IN PORT UNTIL WINDS AND SEAS SUBSIDE.

REPEATING THE 10 AM CDT POSITION . . . 28.4 N . . . 88.0 W. MOVEMENT TOWARD . . . NORTHWEST NEAR 8 MPH. MAXIMUM SUSTAINED WINDS . . . 110 MPH. MINIMUM CENTRAL PRESSURE . . . 963 MB.

AN INTERMEDIATE ADVISORY WILL BE ISSUED BY THE NATIONAL HURRICANE CENTER AT 1 PM CDT FOLLOWED BY THE NEXT COMPLETE ADVISORY AT 4 PM CDT.

GUINEY

STRIKE PROBABILITIES ASSOCIATED WITH THIS
ADVISORY NUMBER CAN BE FOUND UNDER AFOS
HEADER MIASPFAT2 AND WMO HEADER WINT72
KNHC.

Discussion

The discussion explains the hurricane specialist's reasoning for
the predicted track and intensity (in this case, the hurricane specialist's last name is Guiney). Typically included are the forecaster's interpretations of the computer models and current
oceanic observations from ships, buoys, oil rigs, and the reconnaissance flights by the Hurricane Hunters and the Hurricane Research Division.

HURRICANE GEORGES DISCUSSION NUMBER   49
NATIONAL WEATHER SERVICE MIAMI FL
11 AM EDT SUN SEP 27 1998

THE AIR FORCE RESERVE HURRICANE HUNTER
AIRCRAFT RECENTLY REPORTED A MINIMUM
CENTRAL PRESSURE OF 963 MB FROM A GPS
DROPSONDE WITH MAXIMUM FLIGHT-LEVEL WINDS
OF 86 KNOTS. BUOY 42040 REPORTED A 45 KT 8-
MINUTE SUSTAINED WIND . . . APPROXIMATELY 55 KT
1-MINUTE WIND SPEED . . . AND 33 FOOT SEAS AT 12Z
WHILE THE C-MAN BUOY NEAR THE MOUTH OF THE
MISSISSIPPI RIVER WAS 39 KT. THE INITIAL INTENSITY
WILL BE HELD AT 95 KNOTS. THIS KEEPS GEORGES ON
THE HIGH END OF CATEGORY TWO STATUS. THE
FORECAST STILL CALLS FOR GEORGES TO
STRENGTHEN TO A CATEGORY THREE BEFORE
LANDFALL.

BASED ON THE WIND DATA ABOVE . . . RECON WIND
PLOTS AND HURRICANE RESEARCH DIVISION . . .
HRD . . . ANALYSES THE WIND RADII HAVE BEEN
ADJUSTED. THE PRIMARY CHANGE IS THE ADDITION
OF HURRICANE-FORCE WINDS IN THE SOUTHWEST
QUADRANT.

FIXES FROM RECON AND NATIONAL WEATHER
SERVICE DOPPLER RADAR SUGGEST A SLIGHT
SLOWING IN THE FORWARD SPEED OVER THE LAST 6
HOURS. THE INITIAL MOTION ESTIMATE IS 315/07.
THE SYNOPTIC REASONING REGARDING THE TRACK
REMAINS UNCHANGED FROM THE LAST SEVERAL
ADVISORIES. A GRADUAL SLOWDOWN IS FORECAST
AS STEERING CURRENTS WEAKEN. THUS . . . THE
OFFICIAL FORECAST TRACK IS VERY SIMILAR TO THE
PREVIOUS ADVISORY. THIS MEANS THAT THE
HURRICANE COULD PRODUCE EXTREMELY LARGE
RAINFALL AMOUNTS COMBINED WITH A LONG
PERIOD OF ONSHORE WINDS AND STORM SURGE
FLOODING. GEORGES IS A VERY SERIOUS THREAT AND
IT COULD BE EVEN WORSE IF THERE IS FURTHER
STRENGTHENING.

GUINEY

FORECAST POSITIONS AND MAX WINDS
INITIAL	27/1500Z	28.4N	88.0W	95 KTS
12HR VT	28/0000Z	28.9N	88.7W	100 KTS
24HR VT	28/1200Z	29.6N	89.5W	100 KTS
36HR VT	29/0000Z	30.3N	89.9W	80 KTS . . . INLAND
48HR VT	29/1200Z	30.8N	89.9W	65 KTS . . . INLAND
72HR VT	30/1200Z	31.5N	89.5W	50 KTS . . . INLAND

Strike Probability

A strike probability shows the chances of the hurricane center passing within 65 nautical miles of several marine locations in the next 72 hours and is a means to quantify the likelihood of a hurricane passing near a particular coastal location. NHC issues a strike probability discussion every six hours. The interpretation of these numbers requires some detailed explanation, which will begin with an example.

Suppose a hurricane is located in the central Gulf of Mexico, and is forecast to make landfall in southeast Louisiana in 24 hours. Now suppose we could find 100 past cases where a storm was located near this position and was forecast to move along the same track. Since most 24-hour forecasts contain some error, most 24-hour positions will be different than what was predicted. In some cases, the location may be near southeast Louisiana, but in most other cases the actual 24-hour position may be in Texas, Mississippi, Alabama, Florida, or some other location. Now suppose, of these 100 past cases, southeast Louisiana experienced landfall 12 times within 24 hours. Then, the 24-hour strike probability for this hurricane (currently located in the central Gulf of Mexico) making landfall in southeast Louisiana is 12 percent. Likewise, suppose 6 previous hurricanes actually made landfall in Galveston, Texas within 24 hours, then the strike probability for Galveston is 6 percent. In this manner, strike probabilities can be issued for the entire threatened coastline. Similar probabilities can be computed for later forecast times.

The procedure through which NHC computes strike probabilities is somewhat more complicated, but this example conveys the general procedure. A strike probability is defined by NHC as the chance of the hurricane center passing within 65 nautical miles of a particular marine location. These probabilities are based on NHC forecast error statistics for the last 40 most similar cases (in terms of location, date, intensity, and motion), and are an expression of forecast uncertainty. For example, if NHC forecasts a hurricane to be 65 nautical miles from Galveston within 24 hours, what is the probability of this actually happening based on past forecast performance on similar hurricanes located near the same initial location at the same time of year?

NHC issues strike probabilities for the following successive time periods: 1) less than 24 hours; 2) 24–36 hours; 3) 36–48 hours; and 4) 48–72 hours. Since forecast errors increase for longer term forecasts, strike probabilities tend to "cluster" around a particular region for forecasts less than 24 hours, then tend to spread out with incrementally longer forecast time periods. These forecasts are issued in tabular form as shown below. The first column gives the strike probability within 24 hours. The next column gives the added increment to the probability within 36 hours, the next column the added increment within 48 hours, etc. For example, the table below shows that Mobile, Alabama's strike probability within 24 hours is 23 percent, within 36 hours is an added increment of 2 percent (or 25 percent total probability within 36 hours), within 48 hours is an added increment of 1 percent (or 26 percent total probability within 48 hours), and within 72 hours an added increment of 1 percent (or 27 percent total probability within 72 hours). The last column shows the total strike probability within 72 hours. It is important to remember that column 2 through 4 cannot be utilized alone, but as added increments to previous columns.

ZCZC MIASPFAT2 ALL
TTAA00 KNHC DDHHMM
HURRICANE GEORGES PROBABILITIES NUMBER 49
NATIONAL WEATHER SERVICE MIAMI FL
10 AM CDT SUN SEP 27 1998

PROBABILITIES FOR GUIDANCE IN HURRICANE
PROTECTION PLANNING BY GOVERNMENT AND
DISASTER OFFICIALS

AT 10 AM CDT...1500Z...THE CENTER OF GEORGES WAS
LOCATED NEAR LATITUDE 28.4 NORTH...LONGITUDE
88.0 WEST

CHANCES OF CENTER OF THE HURRICANE PASSING
WITHIN 65 NAUTICAL MILES OF LISTED LOCATIONS
THROUGH 7AM CDT WED SEP 30 1998

LOCATION	A	B	C	D	E
29.6N 89.5W	42	X	X	X	42
30.3N 89.9W	31	1	1	X	33
30.8N 89.9W	24	3	1	1	29
MUAN 219N 850W	X	X	X	2	2
JACKSONVILLE FL	X	X	X	2	2
MARCO ISLAND FL	X	X	X	2	2
FT MEYERS FL	X	X	X	2	2
VENICE FL	X	X	X	3	3
TAMPA FL	X	X	X	3	3
CEDAR KEY FL	X	X	1	3	4
ST MARKS FL	X	1	2	5	8
APALACHICOLA FL	6	3	2	4	15
PANAMA CITY FL	6	3	2	4	15
PENSACOLA FL	18	2	1	2	23
MOBILE AL	23	2	1	1	27

LOCATION	A	B	C	D	E
GULFPORT MS	31	1	X	1	33
BURAS LA	42	X	X	1	43
NEW ORLEANS LA	29	1	1	1	32
NEW IBERIA LA	13	5	2	3	23
PORT ARTHUR TX	1	3	4	5	13
GALVESTON TX	X	1	1	6	8
FREEPORT TX	X	X	1	5	6
PORT O CONNOR TX	X	X	X	3	3
GULF 29N 85W	3	1	3	5	12
GULF 29N 87W	65	X	X	X	65
GULF 28N 89W	99	X	X	X	99
GULF 28N 91W	5	3	3	5	16
GULF 28N 93W	X	1	2	7	10
GULF 28N 95W	X	X	X	5	5
GULF 27N 96W	X	X	X	2	2

COLUMN DEFINITION   PROBABILITIES IN PERCENT
A IS PROBABILITY FROM NOW TO 7AM MON
FOLLOWING ARE ADDITIONAL PROBABILITIES
B FROM 7AM MON TO 7PM MON
C FROM 7PM MON TO 7AM TUE
D FROM 7AM TUE TO 7AM WED
E IS TOTAL PROBABILITY FROM NOW TO 7AM WED
X MEANS LESS THAN ONE PERCENT

GUINEY

So, how should these probabilities be used for coastal residents and emergency preparedness officials? Ultimately, this is a personal decision, but general guidelines follow. First, one must decide the time window within which action must be initiated. This depends on the time it will take to complete evacuation procedures and on the intensity of the storm, since gale force winds could arrive well before landfall in strong hurricanes. For populated coastal regions, decisions must be made with smaller probabilities than for those who can afford to wait for more precise forecasts. Second, the person has to decide on a probability threshold value at which action must be taken. This threshold value depends on the risk the decision maker is willing to take for a given probability. A key factor in this decision would be some determination of the cost of not taking action and then being hit by the storm. Obviously, the decision maker should err on the side of caution.

Final Comments on National Hurricane Center Statements

Because hurricanes change direction quickly, one should focus on the predicted path and not extrapolate from the past track. At the same time, one should realize that predicted paths may contain large errors, and typically even average forecasts beyond two days contain errors of several hundred miles. Furthermore, one should not concentrate on the landfall of the center alone, because strong winds, high surf, and torrential rains may extend far from the center.

References

Carter, T. M., 1983. Probability of Hurricane/Tropical Storm Conditions: A User's Guide for Local Decision Makers. National Oceanic and Atmospheric Administration, National Weather Service, 25 pp.

Lawrence, M., 1999. Personal communication.

Neumann, C., 1999. Personal communication.

Sheets, R. C., 1984. The National Weather Service Hurricane Probability Program. NOAA Technical Report NWS 37, National Oceanic and Atmospheric Administration, National Weather Service, National Hurricane Center, Miami, FL.

Maximum Sustained Wind Speed Relationships

Conversion of 10-Minute Sustained Winds to 1-Minute Sustained Winds

The following empirical relationships exist between the 1-minute maximum sustained wind speeds and central pressure in tropical storms and hurricanes. Separate conversion factors exist for the Atlantic and Pacific Oceans. Caution is advised when using the following table, because these relationships are only approximate.

In general, when averaging is applied over a longer period of time, sustained wind speeds decrease because wind gusts occur in short bursts. In other words, 10-minute sustained winds will be less than 1-minute sustained winds under the same weather conditions. Since different countries use different definitions of sustained winds, comparing hurricane intensity statistics between oceans can be problematic. For example, the Atlantic Ocean and North Pacific Ocean basins use a 1-minute averaging sequence, whereas other ocean basins use 10 minutes.

The following conversion factors may be used to convert from 1-minute sustained

Relationship of Central Pressure to Hurricane Winds

1-Minute Maximum Sustained Winds		Central Pressure (mb)	
(kts)	(mph)	Atlantic	Pacific
30	35	1009	1000
35	40	1005	997
45	52	1000	991
55	63	994	984
65	75	987	976
77	89	979	966
90	104	970	954
102	117	960	941
115	132	948	927
127	146	935	914
140	161	921	898
155	178	906	879
170	196	890	858

Source: Dvorak, V. F., 1975. Tropical Cyclone Intensity Analysis and Forecasting from Satellite Imagery. *Mon. Wea. Rev.,* 103, 420–430.

wind speeds to 10-minute, although caution is advised when using this procedure because it is only approximate, and because it is possible for 1-minute sustained winds to be less than 10-minute sustained winds (Holland 1993).

10-minute winds = 0.871(1-minute sustained winds)
1-minute winds = 1.148(10-minute sustained winds)

Computing Wind Gusts from Sustained Wind Speed

While computing sustained winds over a given time period, short bursts of stronger winds, known as *wind gusts,* will occur. These wind gusts can cause isolated pockets of damage that are worse than the surrounding area, and therefore wind gusts are useful to know. To compute these wind gusts, one can multiply by a *gust factor,* defined by the ratio of peak 2-second winds to the sustained wind. The following gust factors can be used for various exposures at 10-meter height (about 33 feet). Adapted from Holland (1993).

Gust Factors

	Ocean	Flat Grassland	Woods/City
1-minute sustained winds	1.25	1.35	1.65
10-minute sustained winds	1.41	1.56	2.14

Source: Holland, G. J., 1993. In *Global Guide to Tropical Cyclone Forecasting.* World Meteorological Organization Technical Document, WMO/TD No. 560, Tropical Cyclone Programme, Report No. TCP-31, Geneva, Switzerland. Also available at http://www.bom.gov.au.bmrc.

Glossary

aerosonde A small robotic aircraft that measures pressure, temperature, moisture, and wind. It is a light aircraft that is extremely fuel efficient and capable of flying long distances, but possibly sturdy enough to survive severe weather such as hurricanes.

beta effect A theoretical 2–3 mph poleward and westward drift of a hurricane induced by the earth's rotation.

centrifugal force An outward-directed force in rotating flow that occurs because an object in motion wants to remain in a straight line. The sharper the curvature of the flow and/or the faster the rotation, the stronger is the centrifugal force.

cloud seeding The attempted stimulation of cloud growth by introducing artificial ice nuclei (such as silver iodide) into supercooled clouds, thereby converting the supercooled water to ice and promoting cloud development through the release of latent heat of fusion.

computer model A computer program that takes in current weather observations and approximates solutions to complicated equations so as to predict future atmospheric values, such as wind, temperature, and moisture.

267

concentric eyewall cycle   A natural (but temporary) weakening process in which a new eyewall forms outside the original eyewall. The outer eyewall "chokes off" inflow to the inner eyewall, causing it to dissipate. The outer eyewall then propagates inward, replacing the original eyewall.

convergence   A region in the atmosphere where air accumulates.

Coriolis force   An apparent force caused by the earth's rotation that deflects the wind to the right of its intended path in the Northern Hemisphere and to the left in the Southern Hemisphere. As a result, hurricanes rotate counterclockwise in the Northern Hemisphere and clockwise in the Southern Hemisphere.

cyclonic rotation   A counterclockwise rotation in the Northern Hemisphere or a clockwise rotation in the Southern Hemisphere.

Doppler effect   A shift in wavelength of radiation emitted or reflected from an object moving toward or away from the observer. Doppler radar translates the motion of air particles, cloud droplets, and raindrops into wind speed and wind direction measurements.

downbursts   Air accelerated to the surface by heavy rain and spread out at speeds up to 100 mph or faster. They are capable of severe localized destruction

dropsonde   An instrument dropped from a plane that measures the profile of the atmosphere from the aircraft to the ground. This instrument deploys a parachute 10 seconds after ejection and falls at 1,000 feet per minute, measuring pressure, temperature, wind, and moisture.

Dvorak technique   A methodology used to estimate the intensity of a depression, tropical storm, or hurricane solely based on satellite-observed cloud organization and cloud height.

easterly wave   See tropical wave.

El Niño-Southern Oscillation (ENSO)   A 12- to 18-month period during which anomalously warm sea surface temperatures occur in the eastern half of the equatorial Pacific Ocean. El Niño events occur irregularly, about once every 3–7 years. Atlantic hurricane activity is usually suppressed during El Niño seasons. The opposite condition is called La Niña.

eye   A region in the center of a hurricane (and tropical storms near hurricane strength) where the winds are light and skies are clear to partly cloudy.

eyewall   A wall of dense thunderstorms that surrounds the eye of a hurricane.

front   A boundary between two air masses of different temperature and/or moisture properties.

Fujiwhara effect   Interaction of two hurricanes (located less than 850 miles from each other) that orbit cyclonically about a midpoint between them; named after the pioneering experiments of Fujiwhara in 1921.

geosynchronous satellites   A satellite that travels east at the same speed as the rotating earth, enabling the satellite to remain over the same location and provide continuous coverage of that region.

global warming   A theoretical enhancement of the greenhouse effect caused by an increase in carbon dioxide by fossil fuel emissions (such as the output from cars running on gasoline).

greenhouse effect   A heat balance between incoming solar radiation from the sun and outgoing infrared radiation from the earth, regulated by the unique absorbing properties of the earth's atmosphere. Most solar radiation passes through the atmosphere unabsorbed by the air's molecules and warms the earth. The earth emits this heat back in the infrared spectrum, but water vapor and carbon dioxide molecules in the atmosphere absorb the infrared radiation. A portion of this absorbed energy is radiated back to the earth, further warming the surface. In this way, the atmosphere acts as an insulating layer, keeping part of the earth's radiation from rapidly escaping to space. The greenhouse effect is a natural process, and without an atmosphere the earth would be a much colder, unlivable planet with an average surface temperature of $0°$ F.

hurricane   A large mass of organized, oceanic thunderstorms with a complete cyclonic circulation and maximum sustained winds of at least 74 mph somewhere in the storm. Also called typhoons in the Northwest Pacific across 180 degrees E, *chubasco* in the Philippines, severe tropical cyclones in Australia, and severe cyclonic storms in India.

hurricane warning   A warning given when it is likely that an area will experience hurricane conditions within 24 hours.

hurricane watch   A hurricane watch indicates that a hurricane poses a possible threat to an area (generally within 36 hours), and residents of the watch area should begin preparations for hurricane conditions.

ice nuclei   A floating aerosol with a molecular structure similar to ice.

instability   A condition in which saturated air forced upward is less dense than surrounding unsaturated air and therefore accelerates upward, forming towering puffy clouds.

intense hurricane   See major hurricane.

Intertropical Convergence Zone   The area near the equator where Southern Hemisphere and Northern Hemisphere air converges. Typically, it is manifested as patches of thunderstorms circling the globe.

inverted barometer effect   The uplift of water in the center of a hurricane as an adjustment to the low air pressure there, corresponding to 3.9 inches of sea level rise for every 10-mb drop in sea level pressure.

La Niña   A 12- to 18-month period during which anomalously cool sea surface temperatures occur in the eastern half of the equatorial Pacific Ocean. La Niña events occur irregularly, about once every 3–7 years. Atlantic hurricane activity is usually enhanced during La Niña seasons. The opposite condition is called El Niño.

latent heat   Energy transfer conveyed through phase changes of matter. For example, the *latent heat of condensation* is the heat energy released when water vapor (a gaseous state) condenses to a liquid state; *latent heat of evaporation* is the heat energy absorbed by water vapor during the evaporation process; and *latent heat of fusion* is the heat energy released when water freezes into ice.

major hurricane   A hurricane that reaches a maximum sustained wind of at least 111 mph. This constitutes a Category 3 hurricane or higher on the Saffir-Simpson scale. Also called an *intense hurricane*.

mesoscale vortices   Whirling vortices 150–500 feet wide that form at the boundary of the eyewall and eye where there is a tremendous change in wind speed. Updrafts in the eyewall stretch these vortices vertically, making them spin faster and capable of severe localized destruction.

midget   A very small hurricane, typically with tropical storm and hurricane-force winds confined within 140 miles of the storm center.

monsoon trough   Areas where the Intertropical Convergence Zone is displaced 10–20 degrees away from the equator. This occurs in regions where air or water temperature increases away from the equator. The vast majority of genesis cases are associated with monsoon troughs.

multidecadal changes   Changes in weather activity that last 20–30 years. For example, Atlantic major hurricane activity has occurred in multidecadal cycles, with few Category 3 or better between 1900 and 1925, many Category 3 or better hurricanes between 1940 and 1960, and few between 1970 and 1994

numerical weather prediction   Forecasting the weather based on the solutions of mathematical equations by high-speed computers.

polar-orbiting satellites   A satellite following a north-south orbit around the earth's poles, providing pictures centered on different longitudes each hour as the earth rotates underneath them.

pressure   The force per unit area exerted by air molecules on a surface. Conversely, the "weight" of the air above a given area of the earth's surface. Its standard unit of measurement is the millibar (mb), although it is also popularly measured as the height of a column of mercury supported by the atmosphere's weight using an instrument called a *barometer*. Sea-level pressure is normally close to 1013 mb, or 30 inches of mercury.

Project STORMFURY   a government-sponsored attempt in the 1960s to weaken hurricanes by cloud seeding just outside the eyewall to stimulate cloud growth, thereby depriving inflow to the eyewall.

Quasi-Biennial Oscillation (QBO)   An oscillation of equatorial winds between 13 and 15 miles aloft. These winds change direction between westerly and easterly every 12–16 months. Westerly winds are associated with more Atlantic hurricanes than easterly, especially when the wind is westerly at both 13 and 15 miles aloft.

radiosonde   An instrument attached to a balloon that measures the vertical profile of temperature, moisture, pressure, and wind up to 40,000 feet from the ground.

reconnaissance planes   Planes that fly into the hurricane's eye and take critical meteorological measurements. During the hurricane penetration, information about the horizontal wind and temperature structure is transmitted to NHC. Once the plane enters the eye, it deploys a tube of instruments (called a dropsonde) that parachutes downward from flight level to the sea, sending valuable measurements back to NHC.

Saffir-Simpson scale A scale relating a hurricane's central pressure, maximum sustained winds, and storm surge to the possible damage it is capable of inflicting. The scale contains 5 categories increasing numerically with damage, with a Category 1 being a minimal hurricane and a Category 5 being a catastrophic hurricane.

seasonal predictions The process of forecasting whether specific weather conditions will be above normal, normal, or below normal. For example, Dr. Bill Gray issues an annual forecast for the Atlantic hurricane activity.

silver iodide A compound whose molecules consist of one atom of silver and one atom of iodide. Its structure resembles ice crystals, and therefore it is used in cloud seeding.

spiral bands Curved thunderstorm bands that propagate around the circulation of a hurricane.

steering current An atmospheric current, generally somewhere between 5,000 and 15,000 feet above the surface, whose direction best relates to the motion of a hurricane.

storm surge An abnormal rise of the sea along a shore due to a meteorological influence, especially a hurricane. It is officially defined as the difference between the actual water level under the hurricane's influence and the level due to the astronomical tide and wave setup.

storm tide The actual sea level as influenced by the storm surge, astronomical tide, and wave setup. In practice, water level observations during post-hurricane surveys are always storm tides.

supercooled water Water that remains in the liquid phase at temperatures colder than 0°C.

sustained winds The average wind speed over a period of time at roughly 33 feet above the ground. In the Atlantic and Northern Pacific Oceans, this averaging is performed over a 1-minute period, and in other ocean basins, over a 10-minute period.

tornado A rapidly rotating column of air that protrudes from a cumulonimbus cloud in the shape of a funnel or a rope whose circulation is present on the ground.

tornado warning A warning issued when a tornado has been observed by trained people called "weather spotters" or inferred by an instrument called Doppler radar.

tornado watch A watch issued when conditions are favorable for tornado development.

trochoidal motion   Short-term, oscillatory motion of a cyclone center about a mean path.

tropical cyclone   The internationally designated general term for all large cyclonically rotating thunderstorm complexes over tropical oceans. It includes depressions, tropical storms, and hurricanes in addition to other large, tropical, cyclonically rotating thunderstorm complexes that contain distinctly different temperature and organization characteristics, such as monsoon depressions and subtropical cyclones.

tropical depression   A large mass of organized, oceanic thunderstorms with a complete cyclonic circulation and sustained winds everywhere less than 39 mph.

tropical disturbance   A large mass of organized, oceanic thunderstorms that has persisted for 24 hours. Sometimes partial rotation is observed, but this is not required for a system to be designated a tropical disturbance.

tropical storm   A large mass of organized, oceanic thunderstorms with a complete cyclonic circulation and maximum sustained winds between 39 and 73 mph somewhere in the storm. A storm is first given a name at this stage.

tropical wave   A westward moving trough, shaped like an upside down "V" similar to a wave, imbedded in northeasterly winds in the tropics. About 55–75 tropical waves are observed in the Atlantic each year, and 10–25 percent develop into a tropical depression or more. They are also called easterly waves.

trough   An elongated area of low pressure.

typhoon   A hurricane that forms over the Western Pacific Ocean west of 180 degrees E.

vertical wind shear   The difference between wind speed and wind direction at 40,000 feet and the surface. Hurricane formation and intensification is favored in regions where the wind is roughly the same speed and blowing from the same direction at all height levels in the atmosphere (known as weak vertical wind shear). Weak wind shear is a favorable condition for the development or maintenance of tropical disturbances, tropical storms, and hurricanes, since their thunderstorm structure will remain intact.

Index

Note: t. indicates table.

Patrick J. Fitzpatrick is an assistant professor of meteorology at Jackson State University in Jackson, Mississippi. His specialties include hurricanes, weather forecasting, and high performance computer applications in meteorology. Fitzpatrick received both his B.S. and M.S. in meteorology from Texas A&M University, and a Ph.D. in the atmospheric sciences from Colorado State University, under the supervision of hurricane expert Dr. Bill Gray.